Patrick Stuedi

Mobile Ad Hoc Networks from Theory to Practice

Patrick Stuedi

Mobile Ad Hoc Networks from Theory to Practice

Fundamental Properties and Services

Südwestdeutscher Verlag für Hochschulschriften

Impressum/Imprint (nur für Deutschland/ only for Germany)

Bibliografische Information der Deutschen Nationalbibliothek: Die Deutsche Nationalbibliothek verzeichnet diese Publikation in der Deutschen Nationalbibliografie; detaillierte bibliografische Daten sind im Internet über http://dnb.d-nb.de abrufbar.

Alle in diesem Buch genannten Marken und Produktnamen unterliegen warenzeichen-, marken- oder patentrechtlichem Schutz bzw. sind Warenzeichen oder eingetragene Warenzeichen der jeweiligen Inhaber. Die Wiedergabe von Marken, Produktnamen, Gebrauchsnamen, Handelsnamen, Warenbezeichnungen u.s.w. in diesem Werk berechtigt auch ohne besondere Kennzeichnung nicht zu der Annahme, dass solche Namen im Sinne der Warenzeichen- und Markenschutzgesetzgebung als frei zu betrachten wären und daher von jedermann benutzt werden dürften.

Verlag: Südwestdeutscher Verlag für Hochschulschriften Aktiengesellschaft & Co. KG
Dudweiler Landstr. 99, 66123 Saarbrücken, Deutschland
Telefon +49 681 37 20 271-1, Telefax +49 681 37 20 271-0
Email: info@svh-verlag.de
Zugl.: Zurich, ETH, Diss., 2008

Herstellung in Deutschland:
Schaltungsdienst Lange o.H.G., Berlin
Books on Demand GmbH, Norderstedt
Reha GmbH, Saarbrücken
Amazon Distribution GmbH, Leipzig
ISBN: 978-3-8381-0656-4

Imprint (only for USA, GB)

Bibliographic information published by the Deutsche Nationalbibliothek: The Deutsche Nationalbibliothek lists this publication in the Deutsche Nationalbibliografie; detailed bibliographic data are available in the Internet at http://dnb.d-nb.de.

Any brand names and product names mentioned in this book are subject to trademark, brand or patent protection and are trademarks or registered trademarks of their respective holders. The use of brand names, product names, common names, trade names, product descriptions etc. even without a particular marking in this works is in no way to be construed to mean that such names may be regarded as unrestricted in respect of trademark and brand protection legislation and could thus be used by anyone.

Publisher: Südwestdeutscher Verlag für Hochschulschriften Aktiengesellschaft & Co. KG
Dudweiler Landstr. 99, 66123 Saarbrücken, Germany
Phone +49 681 37 20 271-1, Fax +49 681 37 20 271-0
Email: info@svh-verlag.de

Printed in the U.S.A.
Printed in the U.K. by (see last page)
ISBN: 978-3-8381-0656-4

Copyright © 2010 by the author and Südwestdeutscher Verlag für Hochschulschriften Aktiengesellschaft & Co. KG and licensors
All rights reserved. Saarbrücken 2010

Acknowledgements

I want to thank all the people who made this work possible. In particular, I want to thank Gustavo Alonso, Timothy Roscoe, Roger Wattenhofer and Antony Rowstron for their insightful comments and their help. Many thanks go to the people from the Systems Research Group at ETH Zurich for all the lively discussions we had. I also want to thank my family who supported me in many ways. My parents for always being there for me throughout the years when help was needed. Christian, my brother, for challenging me in tennis and keeping my physical and mental fitness balanced. Finally, my deepest thanks belong to Martina. I am deeply indebted to her for all the patience she exercised during many work intensive days and weeks. Her love was my main source of motivation throughout this thesis.

Contents

1   Introduction: From Theory to Practice   7

I   Fundamental Properties   11

2   **Fundamental Properties of Ad Hoc Networks**   13
- 2.1   Overview 13
- 2.2   Approach 14
- 2.3   Related Work 15

3   **Model and Computation Method**   21
- 3.1   Network Model 21
- 3.2   Monte-Carlo Method 22

4   **Connectivity**   25
- 4.1   Constant Transmission Power 25
- 4.2   Adaptive Transmission Power 29
- 4.3   Summary 31

5   **Capacity: Model**   33
- 5.1   Overview 33
- 5.2   Node Relationships 35
- 5.3   Scheduling Algorithms 36
- 5.4   Schedule Graph 39
- 5.5   Throughput Capacity 40
- 5.6   Summary 41

6   **Capacity: Validation**   43
- 6.1   Capacity of static networks 43
- 6.2   Capacity of larger networks 45

7   **Continuing Experiments**   51
- 7.1   Network settings 51
- 7.2   Capacity under different interference models 52
- 7.3   Effects of log-normal shadowing 54
- 7.4   Adaptive Power Assignment 56
- 7.5   Summary 57

8 Distributed Bandwidth Reservation   59
   8.1 Problem Statement 60
   8.2 Reservation Recall and Precision 61
   8.3 Precision and Recall in Random Networks 62
   8.4 End-to-end reservations 65
   8.5 Reservation quality and Loss in random networks 66
   8.6 Summary 68

II System   69

9 The invisible ad hoc network   71

10 Heterogeneous Mobile Ad Hoc Networks   73
   10.1 PolyMAC in MANETs using Bluetooth and 802.11 74
   10.2 Overview 75
   10.3 A MAC Layer Approach 76
   10.4 System Evaluation 82
   10.5 Discussion 85
   10.6 Related Work 86
   10.7 Summary 87

11 MAND: Mobile Ad Hoc Network Directory   89
   11.1 Problem Statement 89
   11.2 System Model 91
   11.3 Routing handler 95
   11.4 Interface Semantics 97
   11.5 Evaluation 100
   11.6 Simulation 104
   11.7 Related Work 107
   11.8 Summary 108

12 Service Discovery and MANET-Internet Integration   111
   12.1 The Service Location Protocol (SLP) 112
   12.2 MANET-SLP 113
   12.3 MANET-Internet integration 114
   12.4 Performance 116
   12.5 Related Work 117
   12.6 Summary 118

13 DOPS: Domain Name and Presence Service   119
   13.1 Standard DNS 119
   13.2 DOPS in the MANET 119
   13.3 Seamless integration with the Internet 121
   13.4 Performance 123
   13.5 Related Work 127
   13.6 Summary 128

14 SIPHoc: Distributed Session Initiation in MANETs — 129
14.1 SIP Overview 129
14.2 SIPHoc in MANETs 130
14.3 Internet-connected MANETs 132
14.4 Case Study: VoIP in MANETs 133
14.5 Related Work 137
14.6 Summary 138

15 Network services in MANETs: A Discussion — 139

16 Social Ad Hoc Networking — 143
16.1 Overview 143
16.2 Architecture 143
16.3 Deployment 145
16.4 Measurements 146
16.5 Summary 150

17 Conclusion — 151

Chapter 1

Introduction: From Theory to Practice

Mobile Ad Hoc Networks (MANETs) combine various extreme aspects of todays computer networks. Communication in the network is wireless. Packets may traverse multiple wireless devices (nodes) towards their destination. The network operates without any pre-existing infrastructure. The nodes in the network are mobile. The devices may have only limited resources.

The combination of those network aspects is unique. There are other forms of wireless networks such as Wireless LANs, but they use a pre-existing infrastructure of base-stations for communication. There are infrastructure-less networks, such as Sensor Networks, but there nodes typically are not mobile. Mobility is supported by cellular networks, such as, e.g., GSM or UMTS. But again, those networks are based on a stationary infrastructure which is not only composed of access points, but also of centralized location servers to determine the position of the devices. There are network forms, such as traditional Peer-to-peer networks, which operate without centralized software components, but those systems rely on the Internet Infrastructure and in most cases they are also not wireless.

Mobile Ad hoc Networks (MANETs) are envisioned as playing a significant role in situations where either no network infrastructure is available or reliance on previously present infrastructure is not desired or not possible. Typical application scenarios of MANETs include conferences or meetings, emergency operations such as disaster rescue, and battlefield communications. Another example application is communication among cars on a highway. There, MANETs might be used to propagate certain alerts like congestion or car accidents. Building and deploying applications is difficult due to the constraints induced by MANETs. For certain types of applications it is even unclear whether their deployment is not hindered by fundamental limitations of the network. This could be because their bandwidth or latency requirements cannot be satisfied, or because the application requires full network connectivity all the time. For other classes of applications, it can be shown that, at least in theory, a successful operation is possible up to a certain network size, but once the number of nodes reaches that size some of the application's requirements can no longer be satisfied. And for a third class of applications, the scaling is no issue, but coming up with suitable software architectures for those applications requires new concepts and mechanisms. The lack of infrastructure for instance, requires applications to be decentralized and demands for new protocol architectures.

There is an ever going debate between theoretical and practical research in ad hoc networks. Theory may lead to statements that are useless in practice or produce methods and algorithms which cannot be applied in a real network. On the other hand, practical systems can simply be seen as implementations of previously established theories. We think that in order to build efficient applications for MANETs, one has to understand the theoretic foundations of those networks and their impact on applications under realistic settings. At the same time, studying theoretic foundations has always to be put into perspective of concrete application scenarios. It is important to actually build real systems and to learn from the experience gained.

In this dissertation we attempt to narrow the gap between theory and practice. In the first part of this dissertation, we study some of the fundamental properties of MANETs, such as network connectivity, throughput capacity and quality of service. Thereby, we put strong emphasis on realistic network models and scenarios that reflect concrete applications. For instance, in contrast to other work studying those properties from an asymptotic perspective, we consider bounded network topologies. Moreover, rather than using graph based communication models, we take effects of shadowing radio propagation and interference into account.

The results from this first part of the thesis helped us to understand what type of applications are feasible in MANETs. For instance, our work on capacity has shown that MANETs will not scale up to thousands of nodes. Rather, we believe that in near future MANETs will consist of up to 100 nodes at maximum. Therefore, one realistic scenario for MANETs is to see them as a natural extension of the Internet. In the second part of this thesis, we focus on how to build systems for MANETs such that they can be used seamlessly by traditional Internet based applications.

As a first step towards this goal, we present a virtual network interface which allows nodes with different media access technologies to be integrated into one single IP based MANET. The virtual interface as presented may play an important role in future networks considering the fact that devices are becoming more and more heterogeneous, particularly with respect to their wireless access technology.

In the following part of the thesis, we present several fully decentralized versions of some fundamental networking services (e.g. SLP, DNS, SIP) and show how those services enable standard Internet-based applications to run transparently in ad hoc networks. We believe that the ability to run Internet-based applications seamlessly in MANETs is the key for ad hoc networking to enter everyday life. The software architecture of each of the services we present in this thesis is centered around *MAND*, a simple but efficient distributed key/value (tuple) store. Storing and searching tuples associated with keys is an important building block of services like DNS or SIP. *MAND* provides a platform for those services to be implemented. The design of *MAND* is driven by the previously explored fundamental limitations of ad hoc networks described in the first part of this dissertation. For instance, one novelty of *MAND* is that the distribution of both tuples and requests happens based on piggybacking onto routing messages that the network uses anyway, thus causing no additional message traffic. In the dissertation, we evaluate *MAND* on a testbed and explore larger setups using simulation.

In the last part of this thesis we present a novel application for social networking. *AdSocial* allows users in an ad hoc network to discover each other based on similar interests. The application also allows users to browse other users profile, or to initiate a chat, video or VoIP session. Similar as other services presented in this thesis, also

AdSocial is using *MAND* to efficiently distribute user information across the network without imposing any additional traffic.

Part I

Fundamental Properties

Chapter 2
Fundamental Properties of Ad Hoc Networks

In the first part of this dissertation, we focus our attention on fundamental properties of ad-hoc networks. In this chapter, we give an overview of the topic and discuss related work.

2.1 Overview

For an ad-hoc network to function properly in the first place it must be connected, or mostly connected. Otherwise the network would consist of scattered isolated islands and could not support networking applications. Secondly, the ad-hoc network must have enough capacity to transport the required amount of data between network nodes. We consider connectivity and capacity as generic network properties, independent of particular protocols or applications. We study parameters that directly and substantially affect the connectivity or the capacity of the network.

Whereas in wireless networks with fixed infrastructure (e.g., cellular telecommunication networks or wireless LANs), it is sufficient that each mobile node has a wireless link to at least one base station, the situation in a decentralized ad hoc network is more complicated. To achieve a fully connected ad hoc network, there must be a wireless multihop path from each mobile node to each other mobile node. The connectivity therefore depends on the number of nodes per unit area (node density) and their radio transmission range. Each single mobile node contributes to the connectivity of the entire network. We regard connectivity to be independent of traffic load. On the physical layer, connectivity between nodes is predicted by the radio model. Whether two connected nodes can communicate with each other at any given moment in time depends of course on interference conditions which are directly related to the traffic load and simultaneous communications between other nodes in the network. Due to interference, communication between two connected nodes may drop to lower speeds or even become impossible at certain times. However, in these cases, we say the the link capacity is reduced, instead of saying that the probability of connectivity between those two nodes is decreased. In other words, we consider interference as a capacity-affecting factor and not as a connectivity issue. Connectivity of ad hoc networks under various different network settings will be discussed in Chapter 4 of this dissertation.

Apart from connectivity, a fundamental feature of a wireless ad hoc network is the rate at which it can transport data. In wireless ad-hoc networks communication between nodes takes place over radio channels. As long as all nodes use the same frequency band for communication, any node-to-node transmission will add to the level of interference experienced by other users, which degrades the throughput capacity of each user. Furthermore, since nodes in ad hoc networks act as relays, they have to use their radio device not only for transmitting *their* data, but also the data from other nodes. This creates additional traffic to the network and degrades the throughput capacity per node further. We discuss throughput capacity in Chapters 5, 6 and 7.

As a third fundamental property, we study the possibility to perform *bandwidth reservations*. Bandwidth reservations are one way to provide quality of service (QoS) to applications. QoS is not a requirement as long as the throughput capacity in the network is high enough. While this might be the case in wired networks, it is certainly not true for ad hoc networks. There, throughput capacity is a rare resource and certain applications that need an assured bit-rate rely on QoS support to protect themselves against best effort traffic. One example of such an application is VoIP. In contrast to other applications, like, e.g., video streaming, VoIP has a sharp threshold for the minimum bandwidth it can operate. If the bandwidth is below the threshold an interactive communication is simply impossible. The situations under which bandwidth reservation is feasible in ad hoc networks and the overhead associated with such an operation is discussed in Chapter 8.

2.2 Approach

There has been a tremendous amount of work studying fundamental properties like connectivity and capacity in ad hoc networks. Most results have either been derived by analysis or using some network simulator. Both approaches have their advantages and disadvantages. Analysis helps to understand the basic characteristics and the asymptotic behavior of a certain property. However, analytical methods often cannot capture the physical conditions of the network. This is because as part of the mathematical analysis the problem has to be simplified by either making assumptions about the size of the network (e.g. infinite number of nodes, no area boundaries, etc.), the radio propagation (e.g. path loss radio propagation) or about the interference calculation (protocol model interference). Studies using a network simulator might be able to take more complicated and realistic network models into account, such as, e.g., log-normal shadowing radio propagation or SINR interference. The drawback of network simulators, however, is that they are very much tailored towards the specific network protocols they are using. Thus, results derived from network simulators do not allow for generic statements.

In this dissertation, we have chosen a novel numerical approach which tries to bridge the gap between mathematical analysis and network simulators. In our approach, we model fundamental properties of ad hoc networks as random variables depending on the node distribution, the communication pattern, the radio propagation, the channel assignment, etc. Expected values of the random variables are then computed using Monte-Carlo Methods. The proposed approach presents a new method to study fundamental properties in situations where pure analytical methods fall short, and protocol specific network simulations are not generic enough. This is of particular evidence against the background of the every increasing computing power of today's hardware. For instance, although the computational costs of our model is $O(n^3)$, we were able to compute all the results in this dissertation within a few hours using a cluster of 64 machines.

2.3 Related Work

Fundamental properties of ad hoc networks, such as, e.g., connectivity, have been studied first around 1989 [1], but the field has recently perceived another boost of research efforts. In this section we will discuss related work in the area connectivity, capacity, scheduling, and quality of service for ad hoc networks.

2.3.1 Connectivity

There has been a vast amount of work on connectivity in ad hoc networks. Gupta and Kumar show in [2] that if the radio transmission range of n nodes uniformly distributed in a disc of unit area is set to $r_c = \sqrt{(ln(n) + c(n))/(\pi n)}$, the resulting wireless multihop network is asymptotically connected with probability one if and only if $c(n) \to +\infty$, where $c(n)$ is a linear function in n. Their study, however, makes several assumptions which are unrealistic in practice. For instance, it is assumed that the received radio signal will turn arbitrarily strong if the sender moves very close to the receiver node. In [3], the authors show that if the signal attenuation function does not have a singularity at the origin and is uniformly bounded, then either the network becomes disconnected or the available data rate per node decreases. In [4] it was shown that connectivity of ad hoc networks can be significantly improved if a sparse network of fixed base stations is added to the network. The practical relevance of this study is, however, questionable, since in most cases ad hoc networks will operate without any pre-existing infrastructure. Traditional connectivity analysis where nodes are distributed in a plane has been extended to three dimensional topologies in [5]. Another approach to study connectivity is to ask for the critical number of neighbors each node must have in order to guarantee connectivity [6].

So far, all the work discussed studies the asymptotic behavior of connectivity. But in practice, ad hoc networks might consist of a limited number of mobile nodes. The problem of finite ad hoc networks has not been addressed very often. Bounds on the connection probability and critical transmission range for a finite ad hoc network were given by [7, 8, 9] and recently by [10].

One limitation which is common to the related work presented so far is that the network is modelled as a random geometric graph. It was shown in [11, 12] that a more accurate modeling of the physical layer is important. Nevertheless, only a few results under more realistic environments are available. Effects of shadowed radio propagation on the packet success probability of a fixed distance link have been analyzed in [13]. In [14], the authors study connectivity by also taking interference into account. In [15], the authors showed that non-deterministic variation of signal power may lead to link asymmetry. IEEE 802.11, the MAC protocol often mentioned in combination with ad hoc networks, allows for data transmission only if there exists a bi-directional connection between the two communicating nodes since data packets need to be acknowledged by the receiving node. Recently, connection probability has been analyzed in a shadow fading model [16] but without considering the asymmetric link problem.

2.3.2 Capacity and Scheduling

Capacity and scheduling issues have been in the focus of research for many years [17, 3, 18, 19]. In contrast to the consensus that accurate physical layer models are important, many recent studies are still based on simplified interference models. In [20] the authors use the protocol model [17] to investigate the interaction between channel assignment and distributed scheduling in multi-channel multi-radio wireless mesh networks. In the protocol model, a transmission from a node u is said to be received successfully by another node v, if no node w closer to the destination node is transmitting simultaneously. Broadcast capacity of multihop wireless networks under protocol interference is studied in [21].

The k-hop interference model is an extension to the traditional protocol model in that it considers all nodes within a hop distance of k from the receiver as interfering nodes. Such a model is studied in [22] to derive bounds for the scheduling complexity. However, using the k-hop neighborhood to approximate interference can be very inaccurate. This will be shown in more detail in chapter 8 later in this thesis.

Another way to model interference is to consider all nodes that lie within a disk of a given radius, centered around a receiver, as interferering nodes. In [23], the authors use such a model to describe an improved packet scheduling algorithm based on virtual coordinates. In general, the disk interference model underestimates the actual interference that is perceived during a wireless transmission. This is shown in chapter 7 in this thesis.

The need for more accurate physical layer models has been recognized by some of the earlier work. Dousse and Thiran have studied bound signal attenuation functions using the physical interference model [3]. In the physical interference model, a communication between two nodes u and v is considered successful if the SINR (Signal to Interference and Noise Ratio) at node v (the receiver) is above a certain threshold. Recently, there has been an increasing interest in the physical interference model. One of the first approaches to apply combined topology control and channel assignment algorithms to SINR-based interference models in multi-hop wireless networks can be found in [24]. Joint congestion control and resource allocation, also under physical interference, have been investigated in [25].

In general, the physical interference model is much harder to study analytically than the protocol model. The problem of scheduling under physical interference was proven to be NP complete in [26]. There are some approximation algorithms for the scheduling problem using the SINR interference model [27, 28]. Those algorithms, however, very much depend on the actual topology of the network. In a recent work Goussevskaia et al. [29] have proposed a scheduling algorithm that schedules links with a constant approximation guarantee, regardless of the topology of the network. The downside of [29] is that the algorithm does produce a short schedule for small and sparse networks (of up to a few hundreds of nodes).

All the studies mentioned so far assume a radio propagation model in which the received signal strength is determined as a direct function of the distance between transmitter and receiver. In [12] it was shown that radio propagation in practice

2.3. RELATED WORK

is asymmetric, which causes many problems on various layers in the network stack. Thus, similar to interference, an accurate modeling of signal propagation is fundamental when computing capacity in wireless networks. Two signal propagation models that are considered to be more realistic are the *Log-Normal Shadowing* (LNS) radio propagation and *Rayleigh* fading [30]. One of the first studies using the LNS model is [13]. There, the authors study effects of shadowed radio propagation on capacity, but without considering multihop networks. In [31], the authors study the capacity of ad hoc networks using different transmission and interference range settings per node. Unfortunately, their work assumes a *disk* interference model which in unrealistic in practice. A similar work was also presented earlier by [32], and more recently by [33].

Algorithms to improve delay and throughput performance combining the physical interference model and Rayleigh fading have been proposed in [34]. However, the random effects occuring as part of the Rayleigh fading make it impossible to find an analytical solution to the problem. Therefore, the authors propose integer linear programming (ILP) to compute approximations for the throughput and the delay. The complexity of the mathematical analysis can be reduced by considering physical interference in the path loss radio propagation model. An asymptotic analysis of capacity in the presence of fading including wireless multihop networks is given in [35]. While asymptotic bounds certainly indicate the generic behavior of ad hoc networks for large number of nodes, they do not give any information on concrete throughput capacity and small networks. Recently, there has been some effort to compute concrete throughput values [36, 37, 38] using ILP. However, their work lacks of a realistic network model. In their studies they use the protocol interference model and and isotropic fading.

2.3.3 Quality of Service

In order to support real-time applications, QoS models like IntServ [39] and DiffServ [40] have been proposed by the IETF for fixed networks. In the DiffServ model, traffic is divided into one best-effort class and a few QoS classes. QoS treatment is provided depending on the priority of the traffic. In the context of Mobile Ad Hoc Networks (MANETs), DiffServ has the advantage that no state information on intermediate nodes has to be maintained and no explicit signaling is needed. But the DiffServ model was originally developed for the Internet backbone where ingress and egress nodes are distinguishable. Ingress nodes typically are responsible for admitting flows and detecting contract violations. In an ad hoc network, every node is a potential sender and therefore an ingress node, so the DiffServ model cannot be applied. In addition, simply dividing the resources into several priority classes can not give any bandwidth guarantees to an individual flow. This is different in an IntServ based approach. Here, exclusive flow-based treatment is provided by reserving and allocating parts of the available bandwidth on each node. One drawback is that IntServ needs to store flow specific information on each node which affects its scalability.

SWAN [41] is based on the DiffServ model. Although SWAN is able to maintain some sort of QoS for admitted flows, it is not a sufficient solution as it treats all real-time traffic equally. Like other DiffServ architectures, it provides no QoS guarantees, but rough approximations.

RSVP [42] is an IntServ flow-based QoS protocol in fixed networks. In RSVP [42], QoS is provided through reservations along the transmission path. The round-trip signaling between sender and receiver builds up a flow-path and reserves the necessary bandwidth according to the flow's profile and the resources available along the path. Unfortunately, in the case of MANETs, where paths are constantly changing, RSVP's reservation mechanism is clearly not adequate. There is some work on improving RSVP to suit last-hop wireless access networks. MRSVP [43] and HMRSVP [44] use excessive reservations in all neighboring cells so that QoS can be maintained in whichever cell the mobile hosts might move to. An alternative solution is to find the nearest-common router (NCR), locally repair the reservation from the NCR and the new access point, and restore the original QoS on the new path. Such schemes are used in LSRVP [45] and [46]. Although, these attempts address some QoS issues in wireless environments, RSVP-similar QoS models cannot entirely cope with the needs of QoS provisioning in mobile ad hoc networks. dRSVP [47] is an RSVP-extended protocol which aims at supporting dynamic QoS in wireless network (including MANETs). By requesting bandwidth over a QoS range instead of a single value, dRSVP provides flexible QoS provisioning in a QoS-varying environment. The limitations of dRSVP are: excessive signaling, lack of an effective path reparation mechanism and slow QoS state setup. INSIGNIA [48] is one of the noteworthy QoS frameworks for MANETs. It adopts efficient in-band signaling to piggyback control information into the IP header of traffic so that resource reservation and QoS treatment can be provided along the flow, without the need of a pre-established path. One important limitation of INSIGNIA is that it can only support two QoS levels, not enough to match the needs of fine-grained adaptive real-time applications.

QoS routing protocols such as CEDAR[49] and others [50, 51, 52, 53, 54, 55, 56] interact with resource management to establish paths through the network that meet end-to-end QoS requirements. This might be problematic because the time scales over which session setup and routing operate are distinct and functionally independent tasks. Also MANET routing protocols should not be burdened with the integration of QoS functionality that may be tailored toward specific QoS models. Integrating QoS in routing might even slow down the routing process itself.

Apart from QoS signalling, people have realized that MANETs demand for a totally different approach in implementing the actual reservation of bandwidth. While in wired networks, bandwidth can be controlled locally at each node, this is no longer true in MANETs. Since in MANETs the medium is shared, bandwidth reservations have to be performed on all nodes interfering with that particular node. Several approaches haves been proposed to tackle that problem. Some of them use a *distributed scheduling* mechanism [57, 58, 59, 60] embedded in the MAC layer with the meaning that reservations are mapped to an equal amount of time slots in the MAC layer and interference is avoided by notifying neighboring nodes to not transmit any data during these slots. However, these approaches rely on time synchronization which is very difficult to achieve in practice, especially under node mobility. In a recent work by Salonidis and Tssuiulas [61], the authors relax the assumption about time synchronization but their approach is restricted to tree topologies. In [62], a QoS routing scheme is proposed that takes neighborhood interference into account. This is done using explicit message exchange with these neighbors which causes an expensive message overhead since some of the contending neighbors may not be located in transmission range and can only be reached through multihop messages. Another approach proposed in [63] avoids additional control messages by dynamically adjusting the contention window size for QoS flows. This approach, however, does not provide

2.3. RELATED WORK

admission control for new flows but rather some sort of conflict resolving after a new flow has already been admitted.

Chapter 3
Model and Preliminaries

We begin our study by presenting the model and the terminology we are using.

3.1 Network Model

3.1.1 Deployment Area

We consider $\mathcal{N}$ to be a set of N nodes uniformly distributed on a square area of side length α, $\mathcal{A}_\alpha := \frac{1}{2}[-\alpha, \alpha]^2$. While in a real-world scenario or a computer simulation the deployment area is always finite, analytical calculations are often considerably simplified by working on $\mathcal{A}_\infty$, thereby avoiding boundary conditions. In this dissertation we particularly study the impact of network boundaries on fundamental properties. To do so, we calculate the quantity in question only over a scope — a square sub-area centered in the deployment area — and indicate the size of the scope in the legend or caption of the corresponding figure. By changing the scope, first from being equal to the network area and then to a small part of the network, we will be able to capture the effects of the network boundary on a certain network property.

3.1.2 Radio Propagation Model

In this dissertation we use the *log-normal shadowing* radio propagation model (LNS) (see [30, 64] for experimental evidence, and [65, 16, 66, 67] for related work using the same model). In the log-normal shadowing model, the reception power ϑ_{sh} in distance r from a node transmitting with signal power p is a random variable defined by

$$\vartheta_{sh}(p, r) = p \cdot (r/r_0)^{-\rho} \cdot 10^{X/10} \tag{3.1}$$

with $r_0 > 0$ being the *reference distance for the antenna far-field*, $\rho > 0$ the *path loss exponent*, and X a normal distributed random variable with zero mean and standard deviation σ (referred to as the *shadowing deviation*) [1]. If the shadowing deviation is equal to zero ($\sigma = 0$), the radio propagation range is a perfect circle (this is also called the deterministic *path loss model*). As the shadowing deviation

---

[1] From a physical point of view the received signal power never exceeds the transmitted power. Hence, Equation 3.1 can hold only for $r > r_0$, while for $r \leq r_0$ the definition $\vartheta_{sh}(p, r) = p$ should be adopted. However, numerically this distinction rarely makes a significant difference.

Table 3.1: Frequently used symbols.

$\mathcal{A}$	Node deployment area
$\mathcal{N}$	Set of nodes
$\mu = \|\mathcal{N}\|/\|\mathcal{A}\|$	Node density
P_n^t	Transmission power of node n
r_0	Antenna far-field reference distance
ρ	Path loss exponent
$\beta \cdot p^*$	Threshold power for radio reception (threshold constant β, ambient noise power p^*)
r_t	Threshold distance
X ($\mathbb{E}[X] = 0, \mathrm{Var}[X] = \sigma^2$)	Normal distributed shadowing random variable
$P_{n \leftarrow n'}$	Reception power (at node n when sending from node n'')
x_n	Position of node n with respect to an area $\mathcal{A}$
$\|x_n - x_{n'}\|$	Distance between two nodes n and n'
$\varphi: \mathbb{R}_{\geq 0} \to [0, 1]$ $\varphi(r) = \Pr[P \geq \beta p^* \mid R = r]$	Connection function

grows, the shape of the transmission range becomes more random and irregular. In particular, the larger the shadowing deviation, the more likely distant nodes gain connection and nearby nodes lose connection. Figure 3.1 shows how the transmission area may look in different settings of the LNS model. Table 3.2 shows some typical values of ρ and σ [30].

3.1.3 Symmetric and Asymmetric Radio Propagation

Each node n in the network is supposed to transmit with a signal power $P_n^t \in [0, \infty[$. The tuple notation (n', n) refers to the transmission from a node n' to a node n and $d(n, n') = |x_n - x_{n'}|$ is the distance between the two nodes. We use $P_{n \leftarrow n'} = \vartheta_{sh}(P_{n'}^t, d(n' - n)) \in [0, P_{n'}^t]$ to denote the power of the received signal at node n due to the transmission (n', n). We distinguish *symmetric* and *asymmetric* signal propagation. In the *symmetric* case, $P_{n \leftarrow n'}$ will always be equal to $P_{n' \leftarrow n}$ if P_n^t is equal to $P_{n'}^t$. In the *asymmetric* case, the received signal power is assumed to be dependent on the direction of the transmission, thus $P_{n' \leftarrow n}$ might differ from $P_{n \leftarrow n'}$. Considering that $P_{n' \leftarrow n}$ and $P_{n \leftarrow n'}$ are random variables, the *asymmetric* signal propagation corresponds to throwing two coins for every pair of nodes, while the *symmetric* case corresponds to throwing one coin for every pair of nodes.

3.2 Monte-Carlo Method

In this dissertation we are using *Monte-Carlo* methods to compute the values for different fundamental properties, like connectivity, capacity, etc. Monte Carlo methods are numerical methods that use random numbers to compute quantities of interest. This is normally done by creating a random variable whose expected value is the

3.2. MONTE-CARLO METHOD

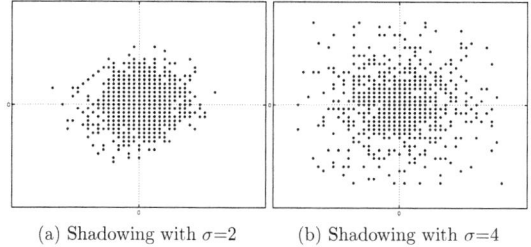

(a) Shadowing with $\sigma=2$ (b) Shadowing with $\sigma=4$

Figure 3.1: Wireless transmission area of a sender at the center of a 400 × 400 grid. Each point marks a position where the signal strength reaches a pre-defined constant threshold

Environment		ρ	σ
Outdoor	Free space	2	4 – 12
	Shadowed/Urban	2.7 – 5	
Indoor	Line-of-sight	1.6 – 1.8	3 – 6
	Obstructed	4 – 6	6.8

Table 3.2: Some typical values of the path loss coefficient (ρ) and the shadowing deviation (σ)

desired quantity. One then simulates the random variable and uses its sample mean to construct probabilistic estimates.

Assume X_n to be the random variable referring to the random position of node n and $X = (X_0, X_1, X_2, \ldots, X_n)$ to be the random vector referring to a random topology. Any given property $prop(X)$, which depends on the random node deployment X, can thus implicitly also be considered as a random variable. The value of interest is the expected value of $prop(X)$, namely $E[prop(X)]$. One could compute $E[prop(X)]$ given the density function $f(prop(X))$ of the random variables $prop(X)$. However, finding the density function with analytical methods is often not possible. But $E[prop(X)]$ can approximately be computed using Monte-Carlo methods:

$$E[prop(X)] = \int_{\mathcal{R}^{2N}} E[prop(X)|X=x]p(X=x)dx$$
$$\approx \frac{1}{k} \sum_{i=0}^{k-1} E[prop(X)|X=x^i] \quad (3.2)$$

Or in other words, we approximately compute the expected value of $prop(X)$ by sampling over k realizations of the underlying random network, with $x^i = (x_0^i, x_1^i, x_2^i, \ldots, x_n^i)$ as a concrete topology deployment.

In the rest of the dissertation, we will use similar methods to approximately compute expected values of specific properties like connectivity or capacity. Those properties may, however, not only depend on the topology X, but also on other random variables like for instance the signal strength distribution or the traffic pattern.

Chapter 4
Connectivity

Most work on connectivity has been done using the deterministic *path loss* model, which predicts the received power as a deterministic function of distance. It is known that the modelling the communication range as an ideal circle is not realistic. In reality, the received power at a certain distance is a random variable due to fading effects. This behavior is reflected by the *shadowing* model, as described in Equation 3.1. In this chapter, we study connectivity in wireless ad hoc networks using the *shadowing* radio propagation model. We first point out how radio irregularity affects the end-to-end connection probability and compare the results to measurements in the path loss model. In a second step, we study the correlation between the *shadowing deviation* (the parameter controlling the level of radio irregularity) and connectivity in more detail and show that the log-normal shadowing model introduces an unnatural bias into the analysis: as the shadowing deviation grows, the radio transmission range not only becomes more irregular, but also enlarges. This naturally leads to an improved connectivity. In a third part, we obtain an unbiased view on the effects of radio irregularity on connectivity in a wireless network. We propose a method to eliminate the bias introduced by the log-normal shadowing model. This allows us to capture the intrinsic properties of radio irregularity and thus to compare different levels of irregularity in a meaningful way. Our approach compensates for the enlarged radio transmission range by adjusting the transmission power of the nodes accordingly. This technique is also used in percolation theory when comparing different network models [68, 69, 70]. Overall, the results presented in this chapter are an important contribution since, when taken together, they show that an analysis of connectivity in a circular radio propagation model yields worst case bounds for connectivity.

4.1 Constant Transmission Power

4.1.1 Direct connection between nodes

Two nodes n and n' can establish a direct communication link (n, n') if the signal strength of the received signal $P_{n \leftarrow n'}$ at node n due to the transmission of node n' is above a certain threshold value, namely $P_{n \leftarrow n'} > \beta \cdot p^*$, where β is the *threshold constant* and p^* the *ambient noise power*.

We define the *threshold distance* as the distance where the received signal power, when $\sigma = 0$, drops to the threshold value $\beta \cdot p^*$. With respect to the shadowing radio propagation model (Equation 3.1), the threshold distance is given by

CHAPTER 4. CONNECTIVITY

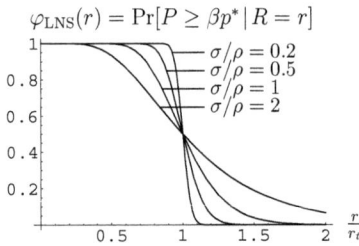

Figure 4.1: Connection function for log-normal shadowing.

$$r_t = r_0 \left(\frac{p_0}{\beta p^*}\right)^{1/\varrho}. \qquad (4.1)$$

The probability that a signal can be received correctly in distance r from a sender node is given by the connection function $\varphi(r)$. From 4.1, the connection function for the log-normal shadowing radio propagation model calculates as[1]

$$\varphi_{\mathrm{LNS}}(r) = \Pr[P \geq \beta p^* \,|\, R = r] = \Pr\left[X \geq \tfrac{10\rho \ln(r/r_t)}{\ln(10)}\right]$$
$$= \tfrac{1}{2} - \tfrac{1}{2}\operatorname{erf}\left(\tfrac{10\rho \ln(r/r_t)}{\sqrt{2}\ln(10)\sigma}\right). \qquad (4.2)$$

The connection function is shown for different values of the shadowing deviation normalized to the path loss exponent (σ/ρ), as the shape depends only on this ratio. For small values of σ/ρ, the connection function becomes a step function and the resulting network graph a unit disk graph with disks of radius $r/r_t = 1$.

4.1.2 Impact of Shadowing on Connectivity

In this section we study how connectivity among nodes over multiple hops evolves with an increasing shadowing deviation. For this, we first define the utility function $\varphi : \mathcal{N} \times \mathcal{N} \to \{0, 1\}$ with

$$\varphi(n, n') = \begin{cases} 1 & P_{n \leftarrow n'} > \beta \cdot p^* \\ 0 & \text{otherwise.} \end{cases} \qquad (4.3)$$

The function $\varphi(n, n')$ computes to one if and only if a successful direct transmission from node n' to node n is possible.

We then say, two arbitrary nodes are connected in the network (possibly though multiple hops), if there is a path between the nodes such that a successful transmission between each two consecutive nodes is possible. Formally, this is described by the connectivity $\zeta_{n,n'} \in \{0, 1\}$ with

---
[1] The error function is defined for all real numbers x as $\operatorname{erf}(x) := \tfrac{2}{\sqrt{\pi}} \int_0^x \exp(-t^2) dt$.

4.1. CONSTANT TRANSMISSION POWER

$$\zeta_{n,n'} = \begin{cases} 1 & \text{there is a path } n_0, n_1, \ldots, n_z, \\ & \text{with } n_0 = n \text{ and } n_z = n' \text{ and} \\ & \varphi(n_{i+1}, n_i) = 1 \text{ for } 0 \geq i < z \\ 0 & \text{otherwise.} \end{cases} \quad (4.4)$$

In principal, $\zeta_{n,n'}$ is a random variable since the position of nodes n and n' is random, as well as the positions and the signal strengths of all other nodes which may or may not participate the path.

We are interested in the expected value of $\zeta_{n,n'}$, for two arbitrary chosen nodes n and n'. This corresponds to the following question. What is the probability of a node to be able to communicate with a randomly chosen destination in the network. Finding the expected value of $\zeta_{n,n'}$ using pure analysis is difficult. Adopting the method described in Section 3.2, $E[\zeta_{n,n'}]$ can approximately be computed as follows.

$$\begin{aligned} \zeta := E[\zeta_{n,n'}] &= \int_{\mathcal{R}^{2N}} E[\zeta_{n,n'}|X=x] p(X=x) dx \\ &\approx \frac{1}{k} \sum_{i=0}^{k-1} E[\zeta_{n,n'}|X=x^i] \\ &= \frac{1}{k} \sum_{i=0}^{k-1} \frac{1}{|\mathcal{N}|} \sum_{n \in \mathcal{N}} \frac{1}{|\mathcal{N}|-1} \sum_{n' \in \mathcal{N} \setminus \{n\}} \zeta_{n,n'}|_{X=x^i} \\ &= \frac{1}{k \cdot |\mathcal{N}| \cdot (|\mathcal{N}|-1)} \sum_{i=0}^{k-1} \sum_{n \in \mathcal{N}} \sum_{n' \in \mathcal{N} \setminus \{n\}} \zeta_{n,n'}|_{X=x^i} \end{aligned} \quad (4.5)$$

Basically, Equation 4.5 computes the connectivity be sampling over k random network topologies and computing the average connectivity level among all pairs of nodes in each deployment[2].

We have computed ζ (Equation 4.5) for different network environments with increasing shadowing deviation σ (see section 3.1.2) and increasing node density. To minimize border effects, we consider only the nodes within a scope while connecting paths may contain intermediate nodes outside of it[3]. We plot the connectivity for different values of the shadowing deviation normalized to the path loss exponent (σ/ρ). Figure 4.2a shows connectivity in the symmetric link model, and Fig. 4.2b for asymmetric links. For symmetric links shadowing improves the connectivity (as observed in [16]). For asymmetric links connectivity decreases as the normalized shadowing deviation grows from 0 to 1. Connectivity increases as the normalized shadowing deviation grows further to 2 and 3. We are not aware that this has been observed before.

---

[2] An alternative definition of connectivity used in literature refers to the probability of the entire network to be connected.

[3] Another option would have been to also exclude nodes outside of the scope from participating the routing. However, we decided to minimize the level modification to not create yet another artificial scenario.

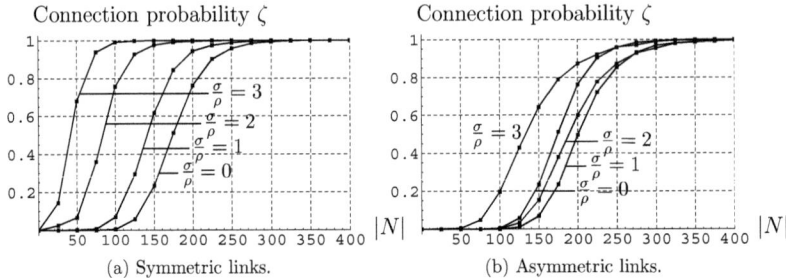

(a) Symmetric links.   (b) Asymmetric links.

Figure 4.2: Connectivity under log-normal shadowing (network settings: $\mathcal{A}_{2000} = [-1000, 1000]^2$ in scope $[-500, 500]^2$, $\rho = 4$, $r_t = 200$).

In order to explain the behavior observed in Figures 4.2a and 4.2b, we study in the next section the expected number of neighbors of a node under different values of the shadowing deviation.

4.1.3 Expected Number of Neighbors

The number of neighbors (or node degree) of a node n can be expressed as follows:

$$neighbors(n) = \sum_{n' \in \mathcal{N} \setminus \{n\}} \varphi(n, n') \qquad (4.6)$$

Again, the expected value $E[neighbors(n)]$ can be approximated by sampling over a large set topology deployments.

$$E[neighbors(n)] = \int_{\mathcal{R}^{2N}} E[neighbors(n)|X = x] p(X = x) dx$$
$$\approx \frac{1}{k} \sum_{i=0}^{k-1} E[neighbors(n)|X = x^i] \qquad (4.7)$$
$$= \frac{1}{k \cdot |\mathcal{N}|} \sum_{i=0}^{k-1} \sum_{n \in \mathcal{N}} neighbors(n)|_{X=x^i}$$

In Fig. 4.3 we plot the expected node degree calculated from Equation 4.7 for different shadowing deviation parameters and different node densities. We do this for the symmetric (Fig. 4.3a) and the asymmetric link model (Fig. 4.3b). Most importantly, the figures capture the bias introduced by the log-normal shadowing radio propagation model: the expected node degree changes with the shadowing deviation. In the symmetric link model, the increase in node degree as shadowing increases naturally improves connectivity. This explains the connectivity increase shown in Figure 4.2a. In the asymmetric link model, the expected node degree first decreases for $0 \leq \sigma/\rho \leq 1.88$, it increases again for $\sigma/\rho > 1.88$. This explains the counterintuitive behavior of connectivity observed in Figure 4.2b.

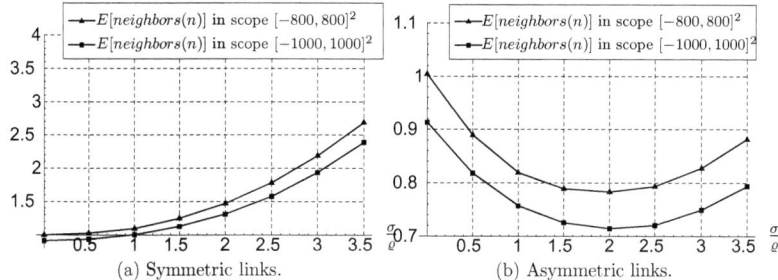

Figure 4.3: Expected node degree under log-normal shadowing (network settings: $\mathcal{A}_{2000} = [-1000, 1000]^2$, $N = 200$, $\rho = 4$, $r_t = 200$).

The figures also illustrate the boundary effects: the results considering the whole deployment area differ quantitatively from those restricted to a smaller scope (as nodes close to the boundary have less neighbors).

These results show that log-normal shadowing primarily affects the node degree, which in turn affects connectivity. This is because an increasing shadowing deviation results not only in a more irregular, but also in an enlarged radio transmission range, which is an unnatural side effect of the log-normal shadowing radio propagation model.

4.2 Adaptive Transmission Power

4.2.1 Connectivity Under Constant Node Degree

To avoid the effect of shadowing enlarging the transmission range we adjust the transmission power of the nodes such that the expected node degree is kept constant for different values of the shadowing deviation. To give some geometric intuition, this is equivalent to keeping the transmission area always the same regardless of the shape (and thus equal to a circle area for $\sigma = 0$).

We can compute the required power values which keeps the expected node degree constant as follows [71].

Corollary 4.1. *In the log-normal shadowing radio propagation model on $\mathcal{A}_\infty = \mathbb{R}^2$, the transmission powers $p_\leftrightarrow$ and $p_\leftrightarrows$ such that*

$$\mathbb{E}[N_\leftrightarrow]|_{p_0 = p_\leftrightarrow} = \mathbb{E}[N_\leftrightarrow]|_{\sigma = 0} \tag{4.8a}$$

$$\mathbb{E}[N_\leftrightarrows]|_{p_0 = p_\leftrightarrows} = \mathbb{E}[N_\leftrightarrows]|_{\sigma = 0} \tag{4.8b}$$

are

$$p_\leftrightarrow = p_0 \exp\left(-\tfrac{(\ln(10)\sigma)^2}{100\varrho}\right) \tag{4.9a}$$

$$p_\leftrightarrows = p_0 \exp\left(-\tfrac{(\ln(10)\sigma)^2}{100\varrho}\right) \left(1 - \operatorname{erf}\left(\tfrac{\ln(10)\sigma}{10\varrho}\right)\right)^{-\tfrac{\varrho}{2}}. \tag{4.9b}$$

In Figure 4.4 we show how connectivity evolves (using Equation 4.5) with the shadowing deviation, this time using the transmission power (Equation 4.9) that

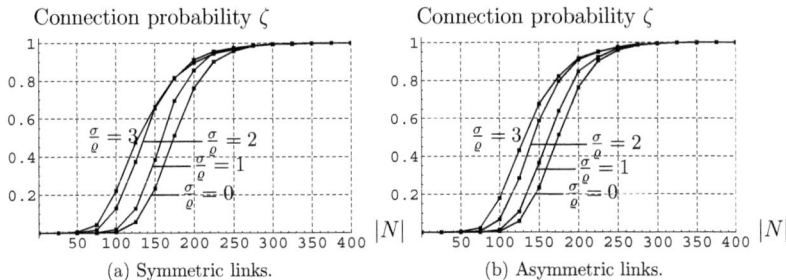

Figure 4.4: Connectivity under log-normal shadowing using the transmission powers 4.9 that preserve the expected node degree (network settings: $\mathcal{A}_{2000} = [-1000, 1000]^2$ in scope $[-500, 500]^2$, $\varrho = 4$, $r_t = 200$).

preserves the expected node degree. Again, we consider both the symmetric (Figure 4.4a) and the asymmetric link model (Figure 4.4b).

Our results demonstrate that *connectivity increases with the shadowing deviation both for symmetric and asymmetric links, if the expected node degree is kept constant.* To the best of our knowledge this is the first result capturing the intrinsic impact of radio irregularity on connectivity under log-normal shadowing. In the next section we discover a plausible reason for the observed phenomenon: an increasing average edge length.

4.2.2 Impact of Shadowing on Edge Length

To explore why connectivity increases with the shadowing deviation, even when the node degree is kept constant, we study the distributions of distance between a randomly chosen pair of nodes (chosen either among all pairs of nodes or pairs that are connected by an edge in the network graph, yielding the edge length distribution of the network graph).

More precisely, we consider the following experiment on a finite deployment area $\mathcal{A}_\alpha = \frac{1}{2}[-\alpha, \alpha]^2$: Randomly choose a pair of nodes *that are connected by an edge* (we assume links to be symmetric throughout this section) and denote by R' the distance between the two nodes. Let $f_{R'}(r)$ be the probability density function of R'.

In Figure 4.5 we plot the edge length distribution $f_{R'}(r)$ for different values of the shadowing deviation (be reminded that they are based on the transmission power of Equation 4.9 that preserves the expected node degree). The figure illustrates how the edge length distribution spreads out with shadowing and increasingly deviates from a triangular distribution. This causes the average edge length to increase. The increase in the average edge length justifies the connectivity improvement caused by radio irregularity as observed in the previous section.

Our results indicate that — contrary to what intuition might suggest — radio irregularity is indeed beneficial for network connectivity.

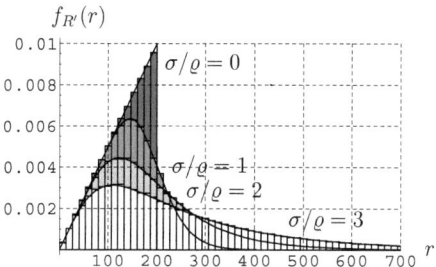

Figure 4.5: Edge length distribution under log-normal shadowing when using the transmission power from 4.9 (network settings: $\mathcal{A}_{16000} = [-8000, 8000]^2$, $N = 12800$, $\varrho = 4$, $r_t = 200$).

4.3 Summary

In this chapter we have studied the impact of log-normal shadowing on connectivity in wireless ad hoc networks.

We have shown that, when using the log-normal shadowing radio propagation model (LNS) to predict the reception power of transmissions in the network, the connectivity increases as the radio signal becomes more irregular. Mainly, this is because the expected node degree increases as the shadowing deviation (the parameter of the LNS model controlling the level of radio irregularity) increases. In order to provide a fair comparison among different levels of radio irregularity, we have computed network connectivity using an adjustment of the transmission power of the nodes that keeps the expected node degree constant. Our results show that even under those circumstances connectivity improves. This is of interest for many existing bounds on the connectivity of wireless networks that have been derived using the deterministic path loss model, as our results indicate these are lower instead of upper bounds on connectivity.

Chapter 5
Capacity: Model

Throughput capacity is another important property of wireless networks. As with connectivity, it is difficult to find closed expressions for the capacity of a wireless network, and even more complicated to analytically study the effects of realistic physical layer models. In this chapter we present a model for throughput capacity that can be used to compute capacity numerically. The approach chosen has a twofold advantage: it is protocol independent and, at the same time, it allows to study effects of various physical layer properties. Experiments using the model will be presented in chapter 7.

5.1 Overview

The main factor that limits capacity in wireless networks is interference, a consequence of using a shared communication medium. An accurate modeling of interference is fundamental in order to derive results of practical relevance. In the literature, two main interference models have been proposed [17]: the *protocol* and the *physical* interference model. In the *protocol* model, a transmission from a node u is said to be received successfully by another node v if no node w closer to the destination node is transmitting simultaneously. However, in practice, nodes outside the interference range of a receiver might still cause enough cumulated interference to prevent the receiver from decoding a message from a given sender. This behavior is captured by the physical model, where a communication between nodes u and v is successful if the SINR (Signal to Interference plus Noise Ratio) at v (the receiver) is above a certain threshold. The physical model can also be less restrictive than the protocol model in a sense that a message from a node u might be correctly received by node v even if there is a simultaneously transmitting node w close to v; for instance because u is using a much larger transmission power than node w. Practical measurements have shown [15, 12] that in some cases a node v might also experience a stronger signal from a node w farther away than another node u, even if both nodes u and w transmit at the same power level. Thus, an accurate modeling of signal propagation is as well important when computing capacity in wireless networks. Similar as done in chapter 4, we use the log-normal shadowing radio propagation (see section 3.1.2) to model the signal propagation in the most realistic way.

Analytical studies on capacity in the presence of realistic radio propagation and interference models – such as log-normal shadowing and physical interference – are

Notation	Semantic
ψ	channel assignment function
κ	interference model
Γ	set of channels
$\mathcal{D}_n$	set of nodes that can be decoded at node n without noise
$\mathcal{U}_n$	set of nodes that can be decoded at node n under noise
$\mathcal{E}$	set of directed edges in a *schedule graph*
$G_T(\mathcal{N}, \mathcal{E})$	schedule graph
$\omega(n', n)$	weight function indicating the number of channels used on a link
Υ	set of source destination pairs
η	routing function
$B_{n',n}$	lowest number of channels between any two neighbors along a path
T	used channels, $T = |\Gamma|$ in $G_T(\mathcal{N}, \mathcal{E})$
$\lambda_{n',n}$	achievable throughput capacity along a path
λ	expected throughput capacity of the network

Table 5.1: Notation

difficult and we are not aware of any existing work. The simplified path loss radio propagation and the protocol interference model are most often used in theoretical studies. These studies typically derive bounds for throughput capacity as the number of nodes in the network tends to infinity. In their seminal work, Gupta and Kumar demonstrated the existence of a global scheduling scheme achieving $\Omega(1/\sqrt{n \log n})$ for a uniform random network with a random traffic pattern [17]. It is not encouraging that the throughput available to each node approaches zero as the number of nodes increases. However, this analysis omits the constant factor that determines whether a realistic and finite network will have a useful per node throughput. While Gupta and Kumar's results illustrate an important property of wireless ad-hoc networks, more concrete statements about capacity in finite networks are necessary to support network planning and deployment.

With this in mind, we adopt a numerical approach based on Monte-Carlo (similar as done for connectivity in chapter 4) methods to study capacity in finite networks and under various interference and radio propagation models, including the physical interference model and log-normal shadowing radio propagation. The approach is centered around a so called *schedule graph* $G_T(\mathcal{N}, \mathcal{E})$ which is directly derived from the physical properties of the network. The effective throughput capacity of a pair of nodes in an ad hoc network is then shown to be related to the connection probability of the corresponding nodes in $G_T(\mathcal{N}, \mathcal{E})$.

In a first step, we want to turn physical properties of wireless multihop networks into a *schedule graph* $G_T(\mathcal{N}, \mathcal{E})$. Examples of physical properties are node locations or perceived signal strengths. In a schedule graph, $\mathcal{N}$ is the set of nodes in the network and $\mathcal{E}$ denotes a set of directed edges between the nodes, such that the existence of a sequence of nodes $n_0, n_1, ... n_k$ – with $n_i \in \mathcal{N}, \forall i \leq k$ and $(n_i, n_{i+1}) \in \mathcal{E}, \forall i < k$ – states that there is also a schedule of channel assignments $\psi(n_0, n_1)$, $\psi(n_1, n_2)$,

...$\psi(n_{k-1}, n_k)$ in a way that node n_0 is able to consecutively transmit data to node n_k at a rate $\lambda_{n_0,n_k} > 0$. The idea behind building a schedule graph is to create an abstraction that allows us to deduce the achievable capacity of the underlying wireless network. In this section, we first define some common properties in order to gradually develop the graph representation. A list of all the notations used within the following two sections, including the aforementioned sets of nodes, can be found in Table 5.1. Throughout this chapter, we use $\mathcal{P}(\cdot)$ to refer to the collection of all possible subsets of a set.

5.2 Node Relationships

Whether a signal from a node n' can be decoded correctly at node n in the absence, or the presence, of concurrent transmissions, is determined by the interference model. In this work, we assume interference models to be defined by a binary interference function $\kappa : \mathcal{N} \times \mathcal{N} \times \mathcal{P}(N) \longrightarrow \{0, 1\}$ with

$$\kappa(n', n, \hat{I}_{n'}) = \begin{cases} 1 & \text{The signal of } n' \text{ can be decoded at} \\ & \text{node } n \text{ under a set } \hat{I}_{n'} \text{ of concurrently} \\ & \text{transmitting nodes} \\ 0 & \text{otherwise.} \end{cases} \quad (5.1)$$

The interference function for the protocol model [17] looks like

$$\kappa_{protocol}(n', n, \hat{I}_{n'}) = 1 \Leftrightarrow d(n'', n) > d(n', n), \forall n'' \in \hat{I}_{n'} \quad (5.2)$$

and for the physical interference model [17] as follows:

$$\kappa_{sinr}(n', n, \hat{I}_{n'}) = 1 \Leftrightarrow \frac{P_{n \leftarrow n'}}{P_n^* + \sum_{n'' \in \hat{I}_{n'}} P_{n \leftarrow n''}} > \beta_{sinr} \quad (5.3)$$

for some threshold β_{sinr} and P_n^* as the thermal noise perceived at node n. Another interference model which is related to the CSMA/CA behavior of 802.11 is the disk interference model. Under disk interference,

$$\kappa_{disk}(n', n, \hat{I}_{n'}) = 1 \Leftrightarrow d(n'', n) > R_I, \forall n'' \in \hat{I}_{n'} \quad (5.4)$$

where R_I is called interference range. The disk model is often used in network simulators such as NS-2 [72]. Typical values for R_I are in between 1.5 and 2.5 times the transmission range. We now assign two sets of nodes to each node $n \in \mathcal{N}$, namely $\mathcal{D}_n$ and $\mathcal{U}_n$.

$$\mathcal{D}_n = \{n' \in \mathcal{N} \mid \kappa(n', n, \emptyset) = 1\} \quad (5.5)$$

is the set of nodes that can be correctly decoded at node n in the absence of any other concurrent transmission.

$$\mathcal{U}_n = \{n' \in \mathcal{D} \mid \kappa(n', n, \hat{I}_{n'}) = 1\} \quad (5.6)$$

contains all nodes n' that can be correctly decoded at node n in the presence of a set of nodes $\hat{I}_{n'}$ transmitting concurrently as node n'. For later use we define $\mathcal{D} = \{(n', n) \mid n' \in \mathcal{D}_n, \forall n \in \mathcal{N}\}$ to be the set of all transmissions in the network when interference is ignored, and $\mathcal{U} = \{(n', n) \mid n' \in \mathcal{D}_n, \forall n \in \mathcal{N}\}$ to be the set of all transmissions in the network if interference is considered.

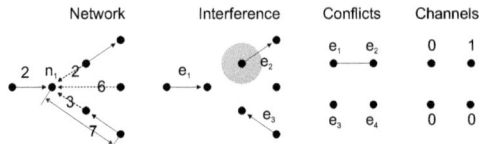

Figure 5.1: Channel assignment under protocol interference.

Algorithm 5.1 Conflict graph under protocol interference
Input: Set of all transmissions $\mathcal{D}$
Output: Set of conflicts $\mathcal{C} \subseteq \{(e, e') \mid e, e' \in \mathcal{D}\}$

1: $\mathcal{C} := \emptyset$;
2: **for all** $e := (n', n) \in \mathcal{D}$ **do**
3:   **for all** $n'' \in \mathcal{N} \setminus \{n, n'\}$ **do**
4:     **if** $d(n'', n) \leq d(n', n)$ **then**
5:       $\mathcal{Q} := \{(n'', n''') \mid n''' \in \mathcal{D}_{n'''}\}$
6:       **for all** $e' \in \mathcal{Q}$ **do**
7:         $\mathcal{C} := \mathcal{C} \cup \{(e', e), (e, e')\}$;
8:       **end for**
9:     **end if**
10:   **end for**
11: **end for**

5.3 Scheduling Algorithms

Which transmissions in the network occur simultaneously is determined by the scheduling algorithm. In our model, we assume the medium to be divided into a set of channels ψ_i. Each channel can be seen as a set of directed transmissions (n', n), with $n' \in \mathcal{D}_n$, between two nodes n' and n. Channels have to be assigned to transmissions in a way such that no two transmissions scheduled within the same channel interfere. At the same time, one wants to keep the number of channels used as low as possible. This is trivial for the protocol and the disk model, but turns out to be more difficult under the physical interference model. In general, the problem of scheduling is related to the traditional graph coloring problem, except that the vertices in the graph to be colored refer to the transmissions in the network and the edges in the graph refer to the interference conflicts. Two vertices conflict if their corresponding transmissions cannot be scheduled simultaneously. We call such a graph a *conflict graph*.

For the protocol model, building a conflict graph is straightforward as shown in Algorithm 5.1. A conflict between two transmissions (n', n), (n''', n'') exists whenever either $d(n''', n) < d(n', n)$ or $d(n', n'') < d(n''', n'')$, where $d(\cdot, \cdot)$ is the distance function $d(n', n) = |x_n - x_{n'}|$. Conflict graph construction and channel assignment under the protocol model are illustrated in Figure 5.1 for a small set of nodes. The straight line arrows in Figure 5.1 represent the transmissions. The numbers assigned to the arrows refer to the distance the two corresponding nodes are apart. According to Algorithm 5.1, the node in the grey area is considered as interfering node. Note that both conflict graph and channel assignment are considered as temporary snapshots from the perspective of e_1.

The procedure is similar in the disk model (Algorithm 5.2), where a conflict

5.3. SCHEDULING ALGORITHMS

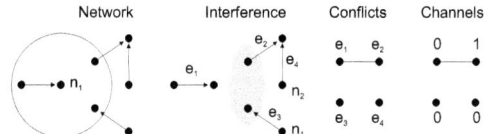

Figure 5.2: Channel assignment under disk interference.

Algorithm 5.2 Conflict graph under disk interference

Input: Set of all transmission $\mathcal{D}$, interference range R_I
Output: Set of conflicts $\mathcal{C} \subseteq \{(e, e^{'}) \mid e, e^{'} \in \mathcal{D}\}$

1: $\mathcal{C} := \emptyset$;
2: **for all** $e := (n^{'}, n) \in \mathcal{D}$ **do**
3:    **for all** $n^{''} \in \mathcal{N} \backslash \{n, n^{'}\}$ **do**
4:      **if** $d(n^{''}, n) \leq R_I$ **then**
5:         $\mathcal{Q} := \{(n^{''}, n^{'''}) \mid n^{'''} \in \mathcal{D}_{n^{'''}}\}$
6:         **for all** $e^{'} \in \mathcal{Q}$ **do**
7:            $\mathcal{C} := \mathcal{C} \cup \{(e^{'}, e), (e, e^{'})\}$;
8:         **end for**
9:      **end if**
10:    **end for**
11: **end for**

---

between two transmissions (n', n), $(n^{'''}, n^{''})$ exists whenever either $d(n^{'''}, n) < R_I$ or $d(n^{''}, n^{'}) < R_I$, where R_I is the interference range. An example of a conflict graph and channel assignment under disk interference is given in Figure 5.2. The straight line arrows in Figure 5.2 represent the transmissions. The disk centered at node n_1 refers to the interference range with radius R_I, compliant with Equation 5.4. According to Algorithm 5.2, nodes in the grey area are considered as interfering nodes. Transmissions e_3 and e_4 do not conflict with e_1 because node n_1 and n_2 are not included in the grey area.

In the physical model, conflicts between two transmissions cannot be determined without considering all other transmissions. As an example, two nodes in distance $2d$ may interfere with a transmission at distance d if their interference is accumulated. For a node $n^{'}$ to belong to $\mathcal{U}_n$, $\kappa_{physical}(n^{'}, n, \mathcal{N} \setminus \mathcal{C}_{n \leftarrow n^{'}})$ must compute to 1, given $\mathcal{C}_{n \leftarrow n^{'}}$ contains all nodes $n^{''}$ acting as a sender in a transmission that conflicts with the transmission $(n^{'}, n)$.

In practice, of course, one wants to find the minimum set $\mathcal{C}_{n \leftarrow n^{'}}$ of conflicts for a transmission $(n^{'}, n)$, because this minimizes the number of channels to be used at a later point in time. How to compute the minimum set of conflicts for a given set of transmissions $\mathcal{D}$ in the physical interference model is shown in Algorithm 5.3. For a given transmission $(n^{'}, n)$, the algorithm operates by gradually testing the SINR ratio with an increasing set of interferer, starting with the node contributing the lowest signal power. At the point where the cumulated interference of a node $n^{''}$ leads to a SINR ratio smaller than β_{sinr}, all transmitting nodes $n^{'''}$ with $P_{n \leftarrow n^{'''}} \geq P_{n \leftarrow n^{''}}$ are considered as interferer and their associated transmissions are defined as conflicts with $(n^{'}, n)$.

We have shown how a conflict graph can be built for the different interference

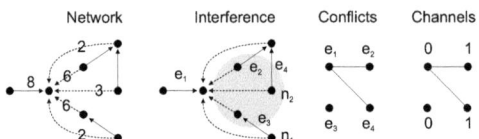

Figure 5.3: Channel assignment under physical interference.

Algorithm 5.3 Conflict graph under physical interference

Input: Set of all transmission $\mathcal{D}$
Output: Set of conflicts $\mathcal{C} \subseteq \{(e, e') \mid e, e' \in \mathcal{D}\}$

1: $\mathcal{C} := \emptyset$;
2: **for all** $e := (n', n) \in \mathcal{D}$ **do**
3: $\quad L := \text{sort}(\mathcal{N} \setminus \{n', n\})$ such that $n'' \prec n''' \longleftrightarrow P_{n \leftarrow n''} < P_{n \leftarrow n'''}$
4: $\quad \mathcal{M}^* := \emptyset$;
5: $\quad$ **for all** $n'' \in L$ **do**
6: $\quad\quad \mathcal{M}^* := \mathcal{M}^* \cup \{n''\}$
7: $\quad\quad$ **if** $\kappa_{sinr}(n', n, \mathcal{M}^*) = 0$ **then**
8: $\quad\quad\quad \mathcal{Q} := \{(n'', n''') \mid n'' \in \mathcal{D}_{n'''}\}$
9: $\quad\quad\quad$ **for all** $e' \in \mathcal{Q}$ **do**
10: $\quad\quad\quad\quad \mathcal{C} := \mathcal{C} \cup \{(e', e), (e, e')\}$;
11: $\quad\quad\quad$ **end for**
12: $\quad\quad$ **end if**
13: $\quad$ **end for**
14: **end for**

models. Based on the conflict graph, efficient coloring algorithms might be used to assign channels to the transmissions (represented as nodes in the conflict graph). Finding the minimum number of channels, however, is an NP hard problem and thus is not feasible for large networks [73]. We decided to apply a *Greedy* channel assignment algorithm (Algorithm 5.4). Algorithm 5.4 assigns channels to transmissions in a greedy way, so that no two transmission e_1, e_2 will be scheduled using the same channel if there exists a conflict between the two transmissions $((e_1, e_2) \in \mathcal{C})$. Algorithm 5.4 further assigns channels in a traffic proportional way, meaning that each node pair (n', n), with $n' \in \mathcal{D}_n$, is assigned exactly as many channels as there are flows occupying the wireless link. Conflict graph and channel assignment for the physical interference model are illustrated in Figure 5.3 in a small example network. The straight line arrows in Figure 5.3 represent the transmissions. The dotted arrows denote signals which contribute to the interference noise of transmission e_1. The weight assigned to an edge corresponds the signal strength. We assume the thermal noise P^* and β_{sinr} used in Equation 5.3 to be 1. According to Algorithm 5.3, nodes in the grey area are considered as the smallest set of nodes such that the remaining cumulated interference does not prohibit transmission e_1 to be established. There is no conflict between transmissions e_1 and e_3 because node n_1 is not included in the grey area. Note that both conflict graph and channel assignment are considered as snapshots from the perspective of e_1.

5.4. SCHEDULE GRAPH

Algorithm 5.4 *Greedy* channel assignment

Input: Set of all transmission $\mathcal{D}$, set of conflicts $\mathcal{C}$
Output: Set of channels $\{\psi_0, \psi_1, ..., \psi_{T-1}\}$ with $\psi_i \subseteq \mathcal{D}$

1: **for all** $e \in \mathcal{D}$ **do**
2:   **for** $i := 0;\ i < \mu(e)$ **do**
3:     $\mathcal{Q} := \{e' \mid (e, e') \in \mathcal{C}\}$
4:     $\Omega := \emptyset;$
5:     **for all** $e' \in \mathcal{Q}$ **do**
6:       $\Omega := \Omega \cup \{\psi_i \mid e' \in \psi_i\}$
7:     **end for**
8:     $\psi_i := freechannel(\Omega);$
9:     $\psi_i := \psi_i \cup \{e\};$
10:   **end for**
11: **end for**

freechannel($\mathcal{Q}$)
1: $\Omega^* := sort(\mathcal{Q})$ such that $\psi_i \prec \psi_j \longleftrightarrow id(\psi_i) < id(\psi_j)$
2: $i := -1$
3: **for all** $\psi \in \Omega$ **do**
4:   **if** $id(\psi) > i + 1$ **then**
5:     **break;**
6:   **end if**
7:   $i := id(\psi);$
8: **end for**
9: **return** $\psi_{i+1};$

5.4 Schedule Graph

Coming back to the definition of $\mathcal{U}_n$, we can say, a node n' belongs to $\mathcal{U}_n$ if $\hat{I}_{n'}$ in Equation 5.6 is defined as the set of nodes transmitting in the same channel as node n'. Given a schedule and the set $\mathcal{U}_n$ for each node, we define a so called schedule graph as a directed and weighted graph $G_T(\mathcal{N}, \mathcal{E})$, where $\mathcal{E}$ denotes the set of directed edges with

$$\mathcal{E} = \{(n', n) \in \mathcal{N} \times \mathcal{N} \mid n' \in \mathcal{U}_n \wedge n \in \mathcal{D}_{n'}\}. \qquad (5.7)$$

The set $\mathcal{E}$ includes all transmissions (n', n) whose signals can be decoded correctly at node n under interference, while the reverse signal might only be correctly decoded if there is no interference[1]. The subscript T indicates the number of channels used (Algorithm 5.4). The weight of an edge $(n', n) \in \mathcal{E}$ is given by $\omega(n', n)$, the transmission capacity function.

$$\omega(n', n) = \sum_{\gamma \in \psi(n', n)} \kappa(n', n, \hat{I}_\gamma). \qquad (5.8)$$

Here, $\hat{I}_\gamma$ denotes the set of nodes transmitting during channel γ, or $\hat{I}_\gamma = \{n' \in \mathcal{N} \mid \gamma \in \psi^*(n')\}$.

It follows directly from the definition of a *schedule graph* $G_T(\mathcal{N}, \mathcal{E})$ that for any path $n_0, n_1, ... n_k$ – with $n_i \in \mathcal{N}, \forall i \leq k$ and $(n_i, n_{i+1}) \in \mathcal{E}, \forall i < k$ – there is also a corresponding schedule of channel assignments $\psi(n_0, n_1), \psi(n_1, n_2), ...\psi(n_{k-1}, n_k)$

---
[1]Note that Equation 5.7 models the acknowledgment as an infinite small packet not occupying the medium

in a way that node n_0 is able to consecutively transmit data to node n_k at a rate strictly greater than zero. We will make use of this property later when deducing the achievable capacity of the underlying physical network.

5.5 Throughput Capacity

Throughout this section, an ad hoc network is represented by its *schedule graph* $G_T(\mathcal{N}, \mathcal{E})$ with the corresponding weight function ω. Capacity is then defined over a given traffic pattern Υ:

$$\Upsilon \subseteq \{(n', n) \in \mathcal{N} \times \mathcal{N} | n' \neq n\}. \tag{5.9}$$

More precisely, we want to know, for a *schedule graph* $G_T(\mathcal{N}, \mathcal{E})$ and a traffic pattern Υ, what is the end-to-end throughput capacity $\lambda_{n',n}$ a given communication pair $(n', n) \in \Upsilon$ will perceive.

Important to the computation of throughput capacity is the routing function $\eta : \mathcal{N} \times \mathcal{N} \longrightarrow \mathcal{P}(\mathcal{E})$. Hence, for a given source-destination pair (n', n) the resulting route simply consists of the set of edges[2] included in the sequence $e_0, e_1...e_{k-1}$, with $e_i = (n_i, n_{i+1}) \in \mathcal{E}$ and $n_0 = n'$ and $n_k = n$.

Now consider the fact that in a *schedule graph*, a path between two nodes also reflects a schedule of channels. Throughput capacity is a concave metric, meaning that the available throughput for a certain source destination pair is always determined by the node with the lowest bandwidth, the so called bottleneck. Hence, let $B^*_{n',n} = \min_{e \in \eta(n',n)} \omega(e)$ be a random variable indicating lowest number of channels available between two nodes along the path from n' to n. One can easily verify that the resulting throughput capacity along the path can not be bigger than $W \cdot B^*_{n',n}/T$, where W is the maximum transmission rate equal to all nodes and $T =| \Gamma |$ is the number of channels used in total.

The throughput capacity may be further diminished when considering all the traffic Υ taking place in the network. For this purpose let us define a so called load function $\mu : \mathcal{E} \longrightarrow [0, N]$, indicating to what extent a certain edge $e \in \mathcal{E}$ is shared among other ongoing traffic, or more formally:

$$\mu(e) = \sum_{\substack{i=\eta(n',n) \\ (n',n) \in \Upsilon}} 1_i(e) \tag{5.10}$$

where $1_i : \mathcal{E} \longrightarrow \{0, 1\}$ is the set membership function. If we want to take all ongoing traffic into account we therefore have to consider μ while computing $B^*_{n',n}$, or

$$B_{n',n} = \begin{cases} 0 & \eta(n', n) = \emptyset \\ \min_{e \in \eta(n',n)} \frac{\omega(e)}{\mu(e)} & \text{otherwise.} \end{cases} \tag{5.11}$$

Based on the definition of $B_{n',n}$ we now claim that the achievable throughput $\lambda_{n',n}$ for a communication pair $(n', n) \in \Upsilon$ in a *schedule graph* $G_T(\mathcal{N}, \mathcal{E})$ can be modelled as

---

[2]Since we assume no loops and the order of the edges in a route is not important for the computation of λ, we prefer the set notion which simplifies further treatment.

5.6. SUMMARY

$$\lambda_{n',n} = \frac{W \cdot B_{n',n}}{T}. \quad (5.12)$$

In a simplified setup, where each node is only allowed to transmit within one single channel, $B_{n',n}$ refers to the path availability between n' and n in $G_T(\mathcal{N}, \mathcal{E})$ and therefore $\lambda_{n',n}$ can be seen as a direct function of the connection probability between the two nodes. Or one can say that the capacity of an ad hoc network is related to the connectivity of its corresponding *schedule graph* $G_T(\mathcal{N}, \mathcal{E})$. This might be of interest when analyzing capacity in sparse and partially disconnected random networks, but also in mobile scenarios where the movement of the nodes often leads to temporarily broken paths.

Since both the network and its graph representation $G_T(\mathcal{N}, \mathcal{E})$ are random, obviously the resulting throughput per node pair $\lambda_{n',n}$ can also be considered as random. Similar as done in chapter 4 we use Monte-Carlo methods to approximately compute the expected value of $E[\lambda_{n',n}]$ of the random variable $\lambda_{n',n}$:

$$\begin{aligned} E[\lambda_{n,n'}] &= \int_{\mathcal{R}^{2N}} E[\lambda_{n,n'} | X = x] p(X = x) dx \\ &\approx \frac{1}{k} \sum_{i=0}^{k-1} E[\lambda_{n,n'} | X = x^i] \\ &= \frac{1}{k} \sum_{i=0}^{k-1} \frac{1}{|\Upsilon|} \sum_{(n,n') \in \Upsilon} \lambda_{n,n'} |_{X=x^i} \\ &= \frac{1}{k \cdot |\Upsilon|} \sum_{i=0}^{k-1} \sum_{(n,n') \in \Upsilon} \lambda_{n,n'} |_{X=x^i} \end{aligned} \quad (5.13)$$

Or in other words, we approximately compute $E[\lambda_{n',n}]$ by sampling over k realizations of the underlying random network while using the mean throughput capacity computed over all communication pairs $(n, n' \in \Upsilon)$. In Equation 5.13 x^i refers to a concrete set of node placements in the area $\mathcal{A}$.

5.6 Summary

In this chapter, we presented a model for studying throughput capacity of wireless multi-hop networks under realistic settings. Contrary to existing work, looking at capacity from an asymptotic perspective based on simplified network models (e.g., protocol interference, unidirectional links, perfect scheduling or straight line routing), our approach analyzes capacity for finite networks under more realistic network configurations. In our model, the effective throughput of a random network is considered as a random variable depending on the node distribution, the communication pattern, the radio propagation, channel assignment, etc..

Chapter 6

Capacity: Validation

In this chapter, we use the model developed in chapter 5 to study capacity under very specific network topologies. We compare our results to results obtained through real measurements and ns-2 simulations [72]. By doing so, we validate our model while at the same time we show the difference between protocol specific throughput capacity (real measurements or simulation based on IEEE802.11) and protocol independent throughput capacity.

6.1 Capacity of static networks

To validate the model, we compute the throughput capacity of two simple, static scenarios, static in the sense that the network topology as well as the communication pattern is fixed. The throughput of such fixed network configurations can be seen as the conditional expected value $E[\lambda_{n,n'}|X=x]$ of the random variable $\lambda_{n,n'}$ under a concrete node placement x. For a fixed channel assignment ψ, we simply use $E[\lambda_{n,n'}|X=x] = \frac{1}{|\Upsilon|}\sum_{(n',n)\in\Upsilon}\lambda_{n,n'}|_{X=x^i}$, where x^i is the given set of coordinates of the nodes. For both examples we will consecutively derive $E[\lambda_{n,n'}|X]$ by going through the basic steps of sections 5.2 – 5.5.

The first network topology we consider consists of three nodes being distance d apart from each other, as shown in Figure 6.1a. To simplify the analysis, we use the more primitive protocol interference model as described in Equation 5.2. Let us further assume $n' \in D_n$ for all $n' \neq n$. According to $\kappa_{protocol}$, the set of senders $\mathcal{U}_n$ is modeled in a way that a node n' belongs to $\mathcal{U}_n$ if, and only if, no other concurrent transmission with a signal stronger than $P_{n\leftarrow n'}$ is received by node n. Hence, the graph G_T only depends on how the different channels are assigned to the nodes. We now want to illustrate the outcome of $E[\lambda_{n,n'}|X=x]$ for three possible channel assignments. We keep track of all states and sets of the network model for each of the three channel assignments in Table 6.1.

In all the configurations we assume shortest path routing and only assign channels to edges that are also used when considering the traffic pattern Υ. In the case of one common channel $\psi(n',n) = \psi(n,n')$ for all nodes n,n', no transmission can correctly be decoded at any receiver (Equation 5.2) which leads to $\mathcal{U}_n = \emptyset, \mathcal{E} = \emptyset$, $S_{n,n'} = 0, \lambda_{n,n'} = 0$ and finally to $E[\lambda_{n,n'}|X=x] = 0$. In the presence of two separate channels ($T = 2$), two directed links can be established (among the potential 6). Along with a communication pattern $\Upsilon = \{(A,B),(B,C),(C,A)\}$ (see Table 6.1),

44   CHAPTER 6. CAPACITY: VALIDATION

T	$\psi(n',n)$	$\mathcal{U}_n$	Υ	$B_{n,n'}$	$E[\lambda\|X]$
1	$\psi(B,A) = \emptyset$ $\psi(C,B) = \emptyset$ $\psi(A,C) = \emptyset$ $\psi(A,B)\{0\}$ $\psi(B,C) = \{0\}$ $\psi(C,A) = \{0\}$	$\emptyset$	(A,B) (B,C) (C,A)	0	0
2	$\psi(B,A) = \emptyset$ $\psi(C,B) = \emptyset$ $\psi(A,C) = \emptyset$ $\psi(A,B) = \psi(C,A) = \{0\}$ $\psi(B,C) = \{1\}$	$\mathcal{U}_A = \emptyset$ $\mathcal{U}_B = \emptyset$ $\mathcal{U}_C = \{B\}$	(A,B) (B,C) (C,A)	$B_{A,B} = 0$ $B_{B,C} = 1$ $B_{C,A} = 0$	1/6W
3	$\psi(B,A) = \emptyset$ $\psi(C,B) = \emptyset$ $\psi(A,C) = \emptyset$ $\psi(A,B) = \{0\}$ $\psi(B,C) = \{1\}$ $\psi(C,A) = \{2\}$	$\mathcal{U}_A = \{C\}$ $\mathcal{U}_B = \{A\}$ $\mathcal{U}_C = \{B\}$	(A,B) (B,C) (C,A)	$B_{A,B} = 1$ $B_{B,C} = 1$ $B_{C,A} = 1$	1/3W
4 4	$\psi(B,A) = \emptyset$ $\psi(C,B) = \emptyset$ $\psi(A,C) = \emptyset$ $\psi(A,B) = \{0,1\}$ $\psi(B,C) = \{2\}$ $\psi(C,A) = \{3\}$	$\mathcal{U}_A = \{C\}$ $\mathcal{U}_B = \{A\}$ $\mathcal{U}_C = \{B\}$	(A,B) (B,C) (C,A)	$B_{A,B} = 2$ $B_{B,C} = 1$ $B_{C,A} = 1$	1/3W

Table 6.1: States for the triangle scenario

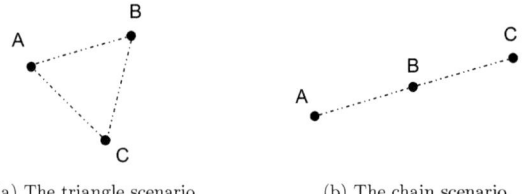

(a) The triangle scenario   (b) The chain scenario

Figure 6.1: Static topologies

$E[\lambda_{n,n'}|X = x]$ is $W \cdot 1/3 \cdot (0 + 1/2 + 0) = 1/6 \cdot W$. If transmissions are spread over three channels, $G_T(\mathcal{N}, \mathcal{E})$ becomes fully connected and $E[\lambda_{n,n'}|X = x]$ equals $W \cdot 1/3 \cdot (1/3 + 1/3 + 1/3) = 1/3 \cdot W$. Adding a fourth edge, e.g., to the transmission between node A and B, does not increase the capacity any further. This is because the increase in the bottleneck ($B_{A,B} = 2$) is compensated by the increase of the total amount of used channels ($T = 4$).

The situation is slightly different for the scenario in Figure 6.1b since node B acts as a router and some of its bandwidth is consumed by traffic sent from A to C. The case $T = 1$ is trivial and comparable with the corresponding case in the

Table 6.2: States for the chain scenario

T	$\psi(n',n)$	$\mathcal{U}_n$	Υ	$B_{n,n'}$	$E[\lambda\|X]$
2	$\psi(A,B) = \psi(C,B) = \{0\}$ $\psi(B,C) = \{1\}$ $\psi(B,A) = \emptyset$	$\mathcal{U}_A = \{B\}$ $\mathcal{U}_B = \emptyset$ $\mathcal{U}_C = \{B\}$	(A,B) (B,C) (C,A)	$B_{A,B} = 0$ $B_{B,C} = 1$ $B_{C,A} = 0$	$1/6W$
3	$\psi(A,B) = \{0\}$ $\psi(B,C) = \{1\}$ $\psi(C,B) = \psi(B,A) = \{2\}$	$\mathcal{U}_A = \{B\}$ $\mathcal{U}_B = \{A\}$ $\mathcal{U}_C = \{B\}$	(A,B) (B,C) (C,A)	$B_{A,B} = 1$ $B_{B,C} = 1$ $B_{C,A} = 0$	$2/9W$
4	$\psi(A,B) = \{0\}$ $\psi(B,A) = \{1\}$ $\psi(B,C) = \{2\}$ $\psi(C,B) = \{3\}$	$\mathcal{U}_A = \{B\}$ $\mathcal{U}_B = \{A,C\}$ $\mathcal{U}_C = \{B\}$	(A,B) (B,C) (C,A)	$B_{A,B} = 1$ $B_{B,C} = 1$ $B_{C,A} = 1$	$1/4W$
			(A,C) (B,C) (C,A)	$B_{A,B} = 1/2$ $B_{B,C} = 1/2$ $B_{C,A} = 1$	$1/6W$
5	$\psi(A,B) = \{0\}$ $\psi(B,A) = \{1\}$ $\psi(B,C) = \{2,3\}$ $\psi(C,B) = \{4\}$	$\mathcal{U}_A = \{B\}$ $\mathcal{U}_B = \{A,C\}$ $\mathcal{U}_C = \{B\}$	(A,C) (B,C) (C,A)	$B_{A,B} = 1$ $B_{B,C} = 1$ $B_{C,A} = 1$	$1/5W$

triangle scenario. Assigning two channels to the four edges results in two established links ($\mathcal{U}_A = \{B\}$ and $\mathcal{U}_C = \{B\}$). Considering the traffic pattern Υ, it is sufficient for one path to be established (e.g., $S_{B,C} = 1$) and the resulting capacity $E[\lambda_{n,n'}|X = x]$ computes to $1/6 \cdot W$. Two of the three routes can be established if 3 channels are used ($T = 3$), which results in $E[\lambda_{n,n'}|X = x] = 2/9 \cdot W$. The capacity of the given traffic pattern can be further improved by assigning one channel per transmission pair. All the routes can be established with a bottleneck of $B_{n,n'} = 1, \forall (n,n') \in \Upsilon$ and $E[\lambda_{n,n'}|X = x]$ equals $1/4 \cdot W$. Obviously, the same channel assignment results in a different capacity if another traffic pattern is used, like, e.g. $\Upsilon = \{(A,C), (B,C), (C,A)\}$. Since the link between node B and C is used twice, the values for $B_{A,C}$ $B_{B,C}$ reduce to $1/2$. For such a traffic pattern, a 5-channel assignment performs better, as shown in Table 6.2.

6.2 Capacity of larger networks

In this section we analyze throughput capacity of various types of communication patterns and network topologies. To simplify the notation we will refer to $E[\lambda_{n,n'}]$ as λ for the rest of the dissertation. For each analyzed configuration we also provide results taken from simulations with ns-2 [72] using the very same topology and communication setup. Throughout this section, we use log-normal shadowing radio propagation with $\sigma = 0$ and $\rho = 2$, and a SINR based interference model, κ_{sinr}, as described in Equation 5.3. Since σ is 0, the threshold for a node n' to be part of $\mathcal{D}_n$ only depends on the distance between the two nodes. We have set P_n^* of

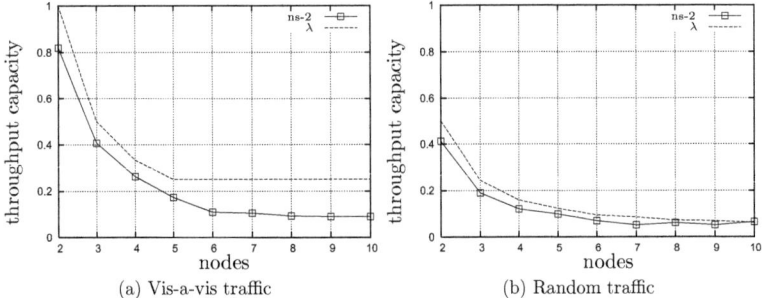

Figure 6.2: Chain topology

Equation 5.3 such that $n' \in \mathcal{D}_n \Leftrightarrow |x_{n'} - x_n| \leq 250$. To avoid mixing up capacity measurements with routing issues, packets within ns-2 simulations are forwarded using pre-computed shortest path routes. We further have set the MAC data rate in ns-2 to 1Mbit/sec. This is necessary since operating 802.11 at higher rates results in drastically reduced efficiency and makes the measurements difficult to compare as the per-packet overhead dominates the overall cost. This is due to the fixed length 802.11 preamble used by the hardware for bit synchronization.

6.2.1 Chain

In a first comparison we look at a configuration of a chain of n nodes. Each node is 200 meters away from its neighbor. The first node acts as a source of data traffic, the last node is the traffic sink. Data is sent as fast as the MAC allows. Since there are no random components involved, λ is a direct function of the channels needed, and computes to $1/4$ as the chain grows. From Figure 6.2a, we see that the value of λ lies above the throughput measured with ns-2, especially when the chain becomes large. This is due to the overhead of headers, RTS, CTS and ACK packets, but also because in reality nodes fail to achieve an optimal schedule. The results obtained with our model match those presented in [19], where the authors discuss throughput capacity measurements taken from ns-2 simulations with respect to theoretical upper bounds.

As a more realistic scenario, we now investigate random communication patterns in chain topologies. For this purpose, we assign a random destination $d(n) \in \mathcal{N}\backslash\{n\}$ to every node $n \in \mathcal{N}$. Figure 6.2b shows the effect of such a traffic pattern on throughput. The plot shows quite a close match between λ and the measurements obtained with ns-2. This is not too surprising since we know from Figure 6.2a that the throughput of an 802.11 chain tallies with the theoretical limit if the chain length is short. Under a random communication pattern the average path length in a chain is far below the maximum value of $n - 1$, for a chain of length n. Furthermore, overlapping communication paths reduce capacity ($B_{n,n'}$ in our model) due to the forwarding load inflicted upon the nodes, especially if the chain becomes long.

6.2. CAPACITY OF LARGER NETWORKS

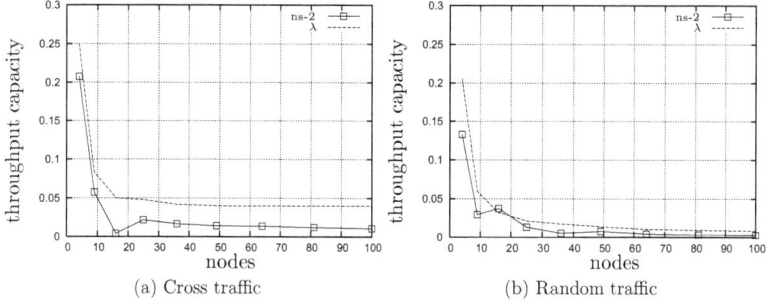

(a) Cross traffic

(b) Random traffic

Figure 6.3: Grid topology

6.2.2 Grid

We look at grid topologies where each node is 200 meters away from its closest neighbor and the nodes communicate using a random communication pattern. Figure 6.3a shows the capacity in the grid topology for a cross communication pattern: source nodes in the first column have a destination assigned in the corresponding row in the last column, and source nodes in the first row have a destination assigned in the corresponding column in the last row. From Figure 6.3a we see that the model based computation predicts a higher throughput capacity than the one measured with ns-2. This is because the cross communication pattern is actually composed of end-to-end chain communications exactly like the scenario used to compute Figure 6.2a. We know that for large chain communications, 802.11 throughput capacity is far below the model-based capacity, for reasons explained in section 6.2.1. In the grid topology with cross communication this behavior is amplified. Figure 6.3b shows the capacity in the grid for a random communication pattern where each node gets assigned a random destination. Similarly to the chain example, the gap between ns-2 measurements and results obtained through our model disappears slightly when communication becomes random. Random communication reduces the average path length and therefore diminishes the impact of the suboptimal channel assignment and the header overhead inherent in 802.11.

6.2.3 Random Topology

We consider random topologies of n nodes distributed uniformly within an area of 1000×1000 square meters. Each node n acts as a traffic generator and has a random destination assigned, chosen uniformly out of $\mathcal{N}\backslash\{n\}$. Figure 6.4a shows the throughput capacity λ in contrast with ns-2 simulation measurements. Similar as in the chain and the grid scenario, the model based prediction matches quite well with the ns-2 simulations. The result supports the trend already observed in the previous configurations of the chain and the grid: the higher the randomness in topology and communication, the better the throughput approximation is. This might particularly be the case in dense networks where the demand for channels is high due to the high node degree, leaving less room for an optimal channel assignment.

48   CHAPTER 6. CAPACITY: VALIDATION

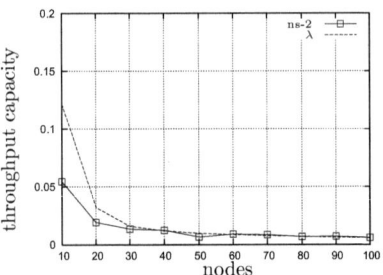

Figure 6.4: Random topology

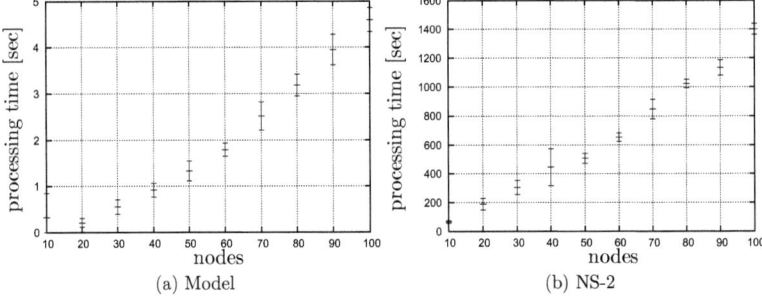

Figure 6.5: Processing time and standard deviation for the throughput capacity of a random network. The time corresponds to the time that was needed to simulate 10 samples of a particular configuration

6.2.4 Summary

In this chapter, we have shown that throughput capacity, computed based on the model proposed in chapter 5, produces reasonable approximations for the potential throughput capacity of arbitrary network configurations. We have validated the model by comparing it to simulation results from ns-2. In general, the simulation results and the model based computations show a similar behavior. In most of the cases, the throughput capacity computed by the model is slightly higher than the one measured with ns-2. This, however, is natural since the ns-2 simulations are based on 802.11 which entails a suboptimal channel assignment and packet overhead. It is however important to note that the model should not be seen as a throughput capacity predictor for $IEEE$802.11 multihop wireless networks, but rather as a tool to study throughput capacity from a protocol independent perspective in situations where pure analysis cannot be used. As one example, we will show in chapter 7 how the model can be used to obtain approximations for throughput capacity under various different physical layer effects, like e.g., shadowing. Another use case of the model are situations where it is interesting to quickly get an estimate of the throughput capacity of a particular network configuration. Using a network simulator like NS-2 often is just too time consuming. Figure 6.5 shows a comparison of the process-

6.2. CAPACITY OF LARGER NETWORKS

ing time and standard deviation needed for both the model based computation and NS-2. For instance, to compute the throughput capacity of a random network with 100 nodes using 10 samples it takes around 1400 seconds ($\approx$ 24 minutes) while the model based computation only takes 4 seconds.

Chapter 7
Continuing Experiments

In this chapter, we use the model developed in chapter 5 to study capacity under various interference and radio propagation models, including the physical interference model and log-normal shadowing radio propagation. Our results shed light into the complex interplay of the different factors influencing throughput capacity: interference, path length, flows per node pair, etc. For instance, we show how shadowing decreases interference as well as the average number of flows between a node pair which leads to a higher throughput capacity.

7.1 Network settings

The network configuration we consider consists of 20 to 300 nodes uniformly distributed on a area of size $2000 \times 2000 m^2$. For the physical interference model, we use a threshold β_{sinr} of 4 decibel, which is the lowest tolerable threshold of an Orinocco PCMCIA Silver/Gold wireless network card so that it can still function at a rate of 1Mbps. The transmission power for every node is kept constant and the thermal noise P^* is adjusted such that the resulting transmission range becomes $200m^2$. For the disk interference model we use $R_I = 400$ (twice the transmission range if $\sigma = 0$). We use Equation 5.13 with a sample size k of 500 to compute the capacity of the network. A random traffic pattern with $|\mathcal{N}|$ flows is considered, where each node is the originator of a flow to a randomly chosen destination, and a shortest path routing is adopted. One important property studied in more detail in the next sections is interference. Remember, for a given transmission (n', n), the nodes $\mathcal{N} \setminus \{(n', n)\}$ can be divided into interfering nodes and non-interfering nodes; the interfering nodes must not transmit concurrently with the transmission (n', n), while the non-interferer may do so. Let's name the set of interfering nodes of a transmission (n', n) as $\mathcal{I}_{n \leftarrow n'}$. How the set $\mathcal{I}_{n \leftarrow n'}$ looks like depends on the interference model used (protocol, physical, disk) and directly follows from the Algorithms 5.1, 5.2 and 5.3. For instance, in the *protocol* interference model, $\mathcal{I}_{n \leftarrow n'}$ contains all nodes n_i with $|x_i - x_n| < |x_n - x_{n'}|$. In the *disk* interference model, $\mathcal{I}_{n \leftarrow n'}$ contains all nodes n_i with $|x_i - x_n| < R_I$, where R_I is the interference range parameter associated with the *disk* model. For the *physical* interference model, $\mathcal{I}_{n \leftarrow n'}$ is the smallest set of nodes that satisfies $\kappa_{sinr}(n, n', \mathcal{I}_{n \leftarrow n'})$. The set of interferer $\mathcal{I}_{n \leftarrow n'}$ can easily be determined by sorting the received signal powers $(P_{n \leftarrow n''})_{n'' \in N \setminus \{n'', n\}}$ and successively adding nodes to the set $\hat{I}$, starting with those that contribute the lowest signal powers, as long as (5.3) is satisfied. Then

$\mathcal{I}_{n;-n'} := (N \setminus \{n', n\}) \setminus \hat{I}$ is the smalles possible set of interferer for the transmission (n', n).
In the following section, we illustrate how capacity, the number of interferer per transmission or the number of flows per node pair, evolve as the network density increases.

7.2 Capacity under different interference models

We first study capacity under different interference models using a regular radio propagation ($\sigma = 0$ in Equation 3.1) and look at effects of irregular radio propagation ($\sigma > 0$ in Equation 3.1) on capacity in the second part of this chapter (Section 7.3).

7.2.1 Interference model

Figure 7.1a shows the capacity of each of the three interference models, *protocol*, *disk* and *physical*. One can clearly observe the gap between the protocol model, which is most commonly used in mathematical analysis, and the physical model, which is supposed to be a more accurate description of a real network.

The two main factors influencing the throughput capacity of a network are interference and the number of flows per node pair (also called *flow-share*). Both factors are directly considered by the capacity model (Equation 5.12 and 5.13) we have proposed in chapter 5. The number of interfering nodes per transmission is shown in Figure 7.1b and the flow-share is shown in Figure 7.1c. As the flow-share is independent of the interference model used, the gap in throughput capacity of Figure 7.1a must results from the difference in the number of interfering nodes per transmission. Figure 7.1b clearly shows that in the physical model the number of interferer per transmission is much higher than under the protocol model. This result contradicts the work in [27], where it is suggested that the physical interference model may increase the network capacity compared to the protocol model.

Figure 7.1e also shows that the disk interference model results in a throughput capacity that is comparable to the one of the physical model with $\beta_{sinr} = 4dB$. This is an important observation since it justifies many of the results produced by the network simulator ns-2, which uses an interference model very similar to the disk interference model we have used.

7.2.2 Capacity in low density networks

The common behavior under all three interference models is that the capacity is high when the network density is low and drops when the network density becomes high. This is the expected behavior since additional nodes introduce more interference and diminish the capacity. The high capacity in the beginning might still be surprising since the network is hardly connected with less than 100 nodes (the connectivity of the network is shown in Figure 7.1d). However, by looking at the numerical data in more detail, we found that in the case of a low network density, some few flows have a very high throughput capacity due to the low interference and the short hop distances. The capacity of these flows compensates the fact that most of the other flows do not even have connectivity and therefore also no capacity. Obviously, although throughput

7.2. CAPACITY UNDER DIFFERENT INTERFERENCE MODELS

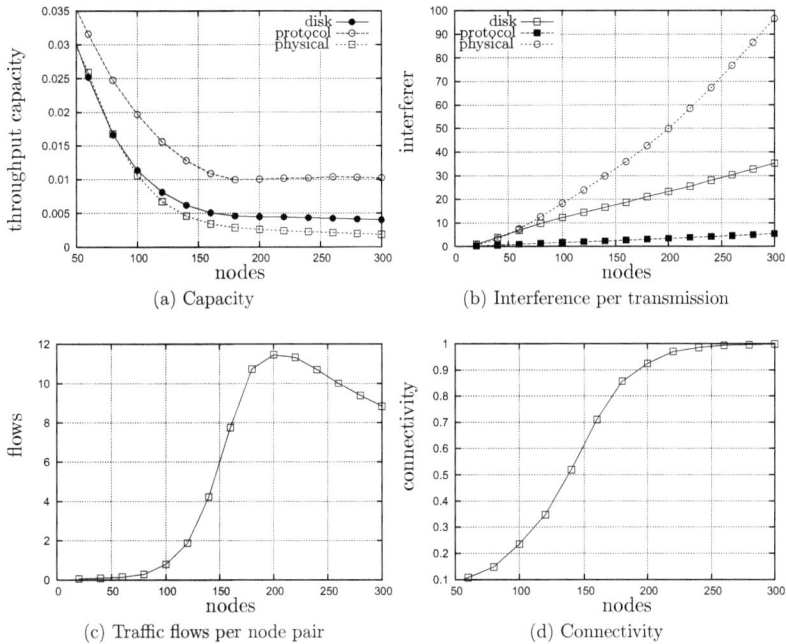

Figure 7.1: Capacity and interference under different interference models (networking settings: $\mathcal{A}_{2000} = [-1000, 1000]^2$, scope $[-700, 700]^2$ $p_0 = 1$, $r_0 = 1$, $\beta = 4$, $r_t = 200$, $\rho = 4$, $\sigma = 0$)

capacity reaches its maximum with a low network density, network planners may not want to consider such settings in practice since the capacity is distributed very unevenly.

7.2.3 Flow-share vs interference

Another interesting observation from Figure 7.1 is that throughput capacity under the protocol interference model seems not to decrease anymore after the network has reached a connectivity level of about 0.9 (according to Figure 7.1c a connectivity of 0.9 is reached with 200 nodes). The explanation lies in the interplay of interference and flow-share. Typically, interference is the dominating factor to determine throughput capacity. However, if the number of interfering nodes becomes small (as in the protocol model), the flow-share starts to become the limiting factor. As one can see from Figure 7.1c, the flow-share first increases up to a level where the network is well connected. Once this level is reached, there are enough alternative paths between any source destination pair and the flow-share is decreasing again. During those high density network configurations, it may happen that the reduction of the flow-share compensates the interference increase and thus additional nodes in the network may even result in a higher throughput capacity.

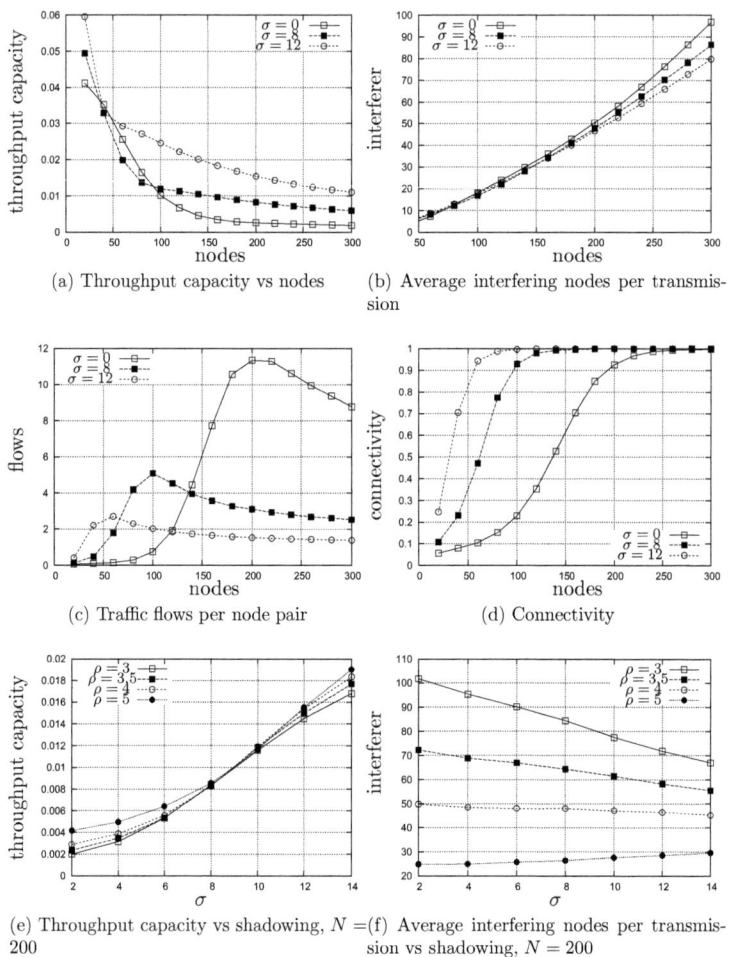

(a) Throughput capacity vs nodes
(b) Average interfering nodes per transmission
(c) Traffic flows per node pair
(d) Connectivity
(e) Throughput capacity vs shadowing, $N = 200$
(f) Average interfering nodes per transmission vs shadowing, $N = 200$

Figure 7.2: Capacity under log-normal shadowing (networking settings: $\mathcal{A}_{2000} = [-1000, 1000]^2$, scope $[-700, 700]^2$, $p_0 = 1$, $r_0 = 1$, $\beta = 4$, $r_t = 200$)

7.3 Effects of log-normal shadowing

We now study irregular radio propagation ($\sigma > 0$ in Equation 3.1) and its impact on throughput capacity. Remember, that the *shadowing deviation* σ can be seen as a measure of radio propagation irregularity. The network configuration for this section corresponds to the settings used in Section 7.2, except that we only consider the physical interference model.

7.3. EFFECTS OF LOG-NORMAL SHADOWING

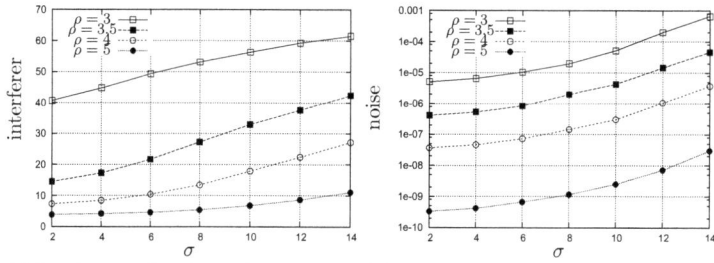

(a) Interfering nodes of a single transmission
(b) Cumulated noise at node n_0. Only nodes outside a circle of radius $r' = 80$ centered at node n_0 are considered

Figure 7.3: Number of interfering nodes and noise perceived of a single fixed-distant transmission (networking settings: $N = 200$, $\mathcal{A}_{2000} = [-1000, 1000]^2$, scope $[-700, 700]^2$, $p_0 = 1$, $r_0 = 1$, $\beta = 4$, $r_t = 200$)

7.3.1 Capacity vs shadowing

We look at throughput capacity for different values of σ. Figure 7.2a and 7.2e illustrate that capacity grows with an increasing value of σ. Similarly as done in section 7.2, we analyze this behavior by looking at interference and flow-share since those are the main factors determining the throughput capacity of a network. As one can observe, both interference (Figures 7.2b and 7.2f) and flow-share (Figure 7.2c) decrease as the shadowing deviation increases. Obviously, and according to Equation 5.12, if interference and flow-share decrease, the throughput capacity of the network must decrease. Therefore, the Figures 7.2b,c,f explain the results shown in Figures 7.2a,e. However, Figures 7.2b,c,f also require explanations themselves. An explanation for the decreasing flow-share (Figure 7.2c) can be found easily. As the shadowing deviation increases the network becomes connected faster (see also Figure 7.2d) and each node may have more neighbors. The increase in neighbors and connectivity naturally leads to more alternative paths between any source destination pair and thus decreases the flow-share. Finding an explanation for the decrease in interference (Figures 7.2b,f) requires a more in-depth analysis.

7.3.2 Interference vs shadowing

It is indeed surprising and counterintuitive to see that interference is decreasing as a side-effect of shadowing. However, please note that the number of interfering nodes shown in Figure 7.2b,f is computed as an average over all node pairs that are participating in communication for a given network configuration and traffic pattern. This is done on purpose since throughput capacity is not affected by the interference of imaginary transmissions of node pairs which do not participate in communication. To see how the number of interfering nodes for a single, fixed distance transmission behaves we have constructed the following artificial experiment. We distribute $N - 2$ nodes uniformly in the network, and additionally placed two nodes, n_0 and n_1, in distance 100 meters from each other at the center of the network area. We fix the

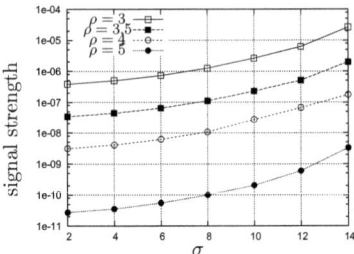

Figure 7.4: Average receiving signal strength of a transmission (networking settings: $N = 200$, $\mathcal{A}_{2000} = [-1000, 1000]^2$, scope $[-700, 700]^2$, $p_0 = 1$, $r_0 = 1$, $\beta = 4$, $r_t = 200$)

shadowing deviation σ to be 0 for the transmission between n_1 and n_0 and leave $\sigma > 0$ for all other transmissions. Figure 7.3a shows that, for such a scenario and for the transmission from node n_1 to node n_0, the number of interfering nodes increases as the shadowing deviation increases. This result is in agreement with the results of chapter 4 where we have shown that the log-normal shadowing radio propagation model increases the transmission range of a node. Indeed, not only the transmission range and interference are increasing in such a scenario, but also the noise (the cumulated signal power of received at a node) that is perceived at node n_0 (Figure 7.3b). Now, when comparing the interference of a single transmission (Figure 7.3a) and interference averaged over all transmissions participating in communication (Figures 7.2b and 7.2f), the following question comes up. Why does shadowing decrease the average interference per transmission while it increase the interference of one single fixed distant transmission? The answer to this question is important since the average interference over all communication pairs is one of the main factors determining the throughput capacity of a network. In order to find the answer we have measured the average received signal power of a transmission that participates the communication pattern. Figure 7.4 shows that the average received signal power increases as the shadowing deviation increases. This makes perfectly sense since, as already mentioned several times, the log-normal shadowing radio propagation model increases the transmission range of a node. Therefore, if the average signal strength of a transmission increases, it takes more nodes to transmit simultaneously and intererfer with the transmission, and thus less nodes have to be considered as interferering nodes. This explains why the average number of interfering nodes per transmission decreases with shadowing (Figures 7.2b,f), while the interference increases for a single fixed-distant transmission (7.3a).

7.4 Adaptive Power Assignment

As shown in the previous section, shadowing using the LNS model decreases the flow-share and the average interference per transmission, and increases the interference of a single fixed distant transmission. Most of those effects are caused by the fact that the transmission range of a node is increasing in the LNS model as the shadowing deviation increases. As discussed in chapter 4, we believe that the increase of the transmission range with shadowing is an artifact of the LNS model and does not

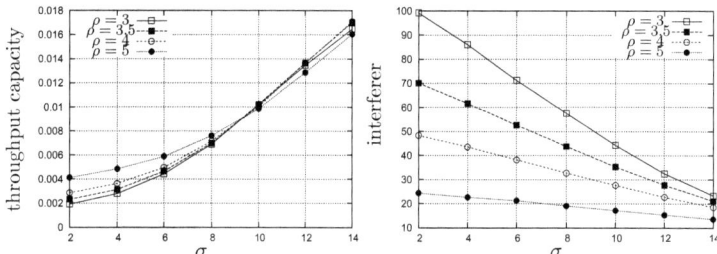

(a) Capacity vs shadowing using an adaptive power assignment

(b) Average interference per transmission using an adaptive power assignment

Figure 7.5: Adaptive power assignment (networking settings: $N = 200$, $\mathcal{A}_{2000} = [-1000, 1000]^2$, scope $[-700, 700]^2$, $p_0 = 1$, $r_0 = 1$, $\beta = 4$, $r_t = 200$, $\rho = 4$)

reflect any real phenomenon. In chapter 4 we have proposed an adaptive power assignment which keeps the expected transmission range (or the expected number of neighbors) of a node constant. We have already shown that such a power assignment improves connectivity. In Figure 7.5 we explore the impact of such an adaptive power assignment on throughput capacity. As one can observe, the throughput capacity still increases, even when using the adaptive power assignment. Figure 7.5b shows that the adaptive power assignment does also not change the way how interference is affected by shadowing. Still the average number of interfering nodes per transmission decreases as shadowing increases.

7.5 Summary

In this chapter, we have studied the impact of various physical layer properties – such as interference or shadowing – on throughput capacity of finite wireless multi-hop networks. As its most important contribution, the results of this chapter show that throughput capacity increases in the LNS model if the shadowing deviation increases. The reason behind it is twofold. First, shadowing in the LNS model improves connectivity and increases the number of neighbors of a node, which creates more opportunities to route traffic through alternative paths rather than enforcing a few node pairs to become the bottleneck. Second, shadowing in the LNS model increases the average signal power of a transmission and thus makes transmissions more resistant against interference. Both effects, shadowing affecting flow-share and interference, are consequences of the fact that shadowing in the LNS model increases the expected transmission range of a node. We have, however, shown in this chapter that even when the transmission range is kept constant, shadowing will still result in a higher throughput capacity of the network. This behavior indicates that previous results based on the circular radio propagation model can be considered as lower bounds instead of upper bounds for throughput capacity.

Chapter 8
Distributed Bandwidth Reservation

In this chapter we want to use some of the mechanisms presented in the previous chapter to explore bandwidth reservation in wireless ad hoc networks. The ability to reserve bandwidth is a key requirement in IntServ based quality of service. An introduction to IntServ in ad hoc networks has been given in chapter 2. In IntServ, QoS is provided through reservations along the transmission path. In MANETs, this is not an easy task due to the need to maintain the reservations in the presence of topology changes and bandwidth variations [43, 44, 47, 48]. In addition, the shared nature of the transmission medium requires reservations to be made not only on the transmission path (*active reservations*) but also on potential interfering nodes (*passive reservations*). As a result, recent work on QoS in MANETs has focused on distributed bandwidth reservation schemas that explore ways to efficiently place active and passive reservations on a MANET. For instance, [57, 58, 59, 60] use a distributed reservation scheme embedded in the MAC layer. Reservations are mapped to an equal amount of time slots at the MAC layer, and interference is avoided by notifying neighboring nodes not to transmit any data during these slots. As another example, [62] proposes a QoS routing scheme that takes neighborhood interference into account. The fundamental underlying problem for such distributed reservation schemas is how to locally determine the set of interfering nodes. Existing work bypasses this problem by using the notion of a k-neighborhood: all nodes within a distance of k hops are considered as interfering nodes.

We apply Monte-Carlo methods to analyze the impact of using the k-neighborhood to identify interfering nodes. We do so by introducing two novel concepts: *reservation precision* (how many nodes where reservations are placed are really interferer) and *reservation recall* (how many nodes where a reservation is needed are actually reserved). We then study how reservation recall and precision evolve as functions of node density, shadowing (radio fading), and path length. To the best of our knowledge, this is the first attempt at quantifying the impact of the use of the k-neighborhood on the efficiency of distributed bandwidth reservation schemas. By itself, this is an important contribution. Yet, the most significant contribution of the analysis is what it shows. Our results point out that there is a inherent trade-off between the quality of a reservation and the amount of resources wasted due to over-reservations. In other words, current approaches can only achieve reasonable QoS guarantees by indulging in severe over-reservation. Considering that in practice ef-

CHAPTER 8. DISTRIBUTED BANDWIDTH RESERVATION

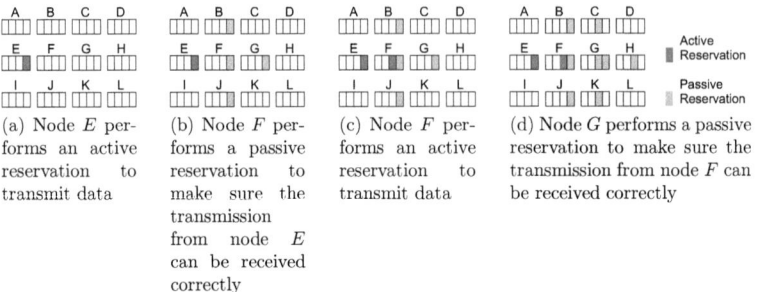

(a) Node E performs an active reservation to transmit data

(b) Node F performs a passive reservation to make sure the transmission from node E can be received correctly

(c) Node F performs an active reservation to transmit data

(d) Node G performs a passive reservation to make sure the transmission from node F can be received correctly

Figure 8.1: Typical reservation process in the 1-hop case

fects like mobility or transmission errors would diminish the reservation quality even further, our analysis makes it evident that existing distributed bandwidth reservation techniques using the k-neighborhood approach are not feasible in MANETs unless traffic is severely restricted.

8.1 Problem Statement

Figure 8.1 shows how a distributed bandwidth allocation process looks when the set of interfering nodes is approximated with a 1-hop neighborhood. The network consists of 12 nodes, labeled from A to L. Every node has a maximum of 4 nodes that are considered as neighbors: the one immediately to its left and right and the one above and below it (e.g., nodes B, E, G, J are neighbors of node F). We assume a simple TDMA-based channel allocation schema. We define an *active reservation* as the set of time slots to be used for transmitting data. A *passive reservation* is the set of time slots required to remain unused in order to not interfere with the transmission. Suppose node E wants to set up a bandwidth reservation for a one-way communication with node G. For simplicity, let the required bandwidth for the entire connection be 1 unit. To begin with, node E locally makes a local 1 unit active reservation to be used for transmitting data to its neighboring node F (In a TDMA-based system this corresponds to a reservation of one time slot). In a second step, node F has to make sure that the reception of the packet from node E is not disturbed by any transmission from interfering nodes. It does so by placing a passive reservation in its 1-hop neighborhood. In a TDMA-based system, this corresponds to informing all the neighbors not to transmit in any of the time slots node E uses to communicate with node F. In addition to the passive reservation, node F makes a 1 unit active reservation to transmit data to node G. Note that node F has both an active and a passive reservation at that time. In a final step, node G places passive reservations in its 1-hop neighborhood to make sure that it is not disturbed by any interfering node while receiving data from node F.

The problem with the reservation process shown in Figure 8.1 is that the 1-hop neighborhood does not match the set of interfering nodes for any node. In fact, wireless interference is complex and nodes far beyond the set of neighbors may actually interfere with a certain transmission. Many protocols use the k-neighborhood, with

8.2. RESERVATION RECALL AND PRECISION

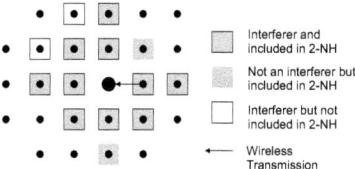

Figure 8.2: A possible arrangement of a 2-hop neighborhood and interfering nodes.

$k \in \{2, 3, 4, ...\}$, as an approximation for the interference area. Obviously, the larger k, the more interfering nodes will be covered, which increases the quality of the reservation. However, the larger k, the more likely it becomes that some of the nodes included in the k-neighborhood do not actually belong to the set of interfering nodes, leading to unnecessary reservations and wasted bandwidth. This is illustrated in Figure 8.2 in the case of a 2-neighborhood and for a possible arrangement of interferer.

The proportion of interfering k-neighbors to all interferer is called *reservation recall*. Reservation recall models the quality of a reservation. Ideally, the reservation recall for a given transmission would be one. This is the case when all interfering nodes are also covered by the k-neighborhood. The proportion of interfering k-neighbors to all k-neighbors is called *reservation precision*. Reservation precision models the amount of resources that are wasted due to a reservation. For instance, in a TDMA-based system, a low reservation precision would mean that most of the reserved slots could actually be used without disturbing any of the ongoing transmissions. Ideally, reservation precision would be one; this is the case if the k-neighborhood contains no nodes that are not interfering. In a perfectly distributed reservation, if the k-neighborhood exactly matches the set of interfering nodes, both reservation recall and reservation precision would be one. We will see that this is hardly the case in random topologies and that optimizing for recall is achieved at the cost of precision, and vice versa.

8.2 Reservation Recall and Precision

Given a link $e = (n', n) \in \mathcal{E}$ and a set of nodes $\mathcal{Q}_{n \leftarrow n'}$ holding passive and active reservations for this link, we would ideally like $\kappa_{sinr}(n', n, \mathcal{N} \setminus \mathcal{Q}_{n \leftarrow n'})$ to be 1. This is the case when $\mathcal{Q}_{n \leftarrow n'}$ exactly matches the set of interferer for the given transmission. For a discussion on how to compute the minimum set of interferer $\mathcal{I}_{n \leftarrow n'}$ for a link $e = (n', n)$ please consider section 7.1 in chapter 7.

Reservation recall and precision are comparisons of the set of interferer with the k-neighborhood. The k-neighborhood is defined to be the set of all nodes that can be reached within k hops, including the node itself. Which nodes can be reached within k hops from a given source node is determined by the routing function η, for which we use the definition of section 5.5. With the routing function η, the k-neighborhood of a node n is

$$\text{k-NH}_n = \{n' \in \mathcal{N} \mid |\eta(n', n)| <= k\}. \tag{8.1}$$

Reservation recall and precision are formally defined on a per edge basis. Reservation recall $R^k_{n \leftarrow n'}$ is the ratio of interfering k-neighbors to all interferer:

$$R^k_{n \leftarrow n'} = \frac{\mid \text{k-NH}_n \cap \mathcal{I}_{n \leftarrow n'} \mid}{\mid \mathcal{I}_{n \leftarrow n'} \mid}. \tag{8.2}$$

Reservation precision $P^k_{n \leftarrow n'}$ is the ratio of interfering k-neighbors to all k-neighbors:

$$P^k_{n \leftarrow n'} = \frac{\mid \text{k-NH}_n \cap \mathcal{I}_{n \leftarrow n'} \mid}{\mid \text{k-NH}_n \mid}. \tag{8.3}$$

Both $R^k_{n \leftarrow n'}$ and $P^k_{n \leftarrow n'}$ are random variables depending on the random node deployment, the signal propagation, etc. Similarly, as done for connectivity (Chapter 4) and capacity (Chapter 5), we use Monte-Carlo methods to approximately compute expected values $E[R^k_{n \leftarrow n'}]$ and $E[P^k_{n \leftarrow n'}]$.

$$E[R^k_{n \leftarrow n'}] = \frac{1}{|\mathcal{E}| \cdot k} \sum_{i=0}^{k-1} \sum_{(n,n') \in \mathcal{E}} R^k_{n \leftarrow n'}|_{X=x^i} \tag{8.4}$$

8.3 Precision and Recall in Random Networks

8.3.1 Network settings

We study reservation recall and precision, R^k and P^k, under different network densities and signal propagation settings. The network configurations we consider consist of randomly deployed nodes within a square of varying size. We use the log-normal radio propagation model (Equation 3.1) and if nothing else is mentioned, the path loss coefficient ρ and the shadowing deviation σ are fixed to be 4 and 0 respectively. For the interference model (Equation 5.3) we use a threshold β_{sinr} of 4 decibel. The transmission power for every node is kept constant and the thermal noise P^* is adjusted in a way that the resulting transmission range becomes $200m^2$. We use Equation 8.4 with a sample size k of 1000[1]. For routing, we use the shortest path algorithm by Floyd and Warshall [74]. Similarly as done in chapter 7, we use a special *scope*, a subset of nodes within a rectangular area, to compute precision and recall.

8.3.2 Recall/Precision Trade-off

As for the first experiment we compute reservation recall and precision for various different values of k and illustrate the result in a scatter-plot (Figure 8.3a). The scatter-plot includes a point for every one of the 1000 samples taken. The x and y coordinates of the dots refer to the corresponding recall and precision values of the sample. The plot includes samples for values of k from 1 to 8. The scatter-plot illustrates the trade-off between reservation recall and precision: a high precision implies a low recall and vice-versa. The result is disappointing since it says that in a network where nodes are deployed randomly, a good reservation quality can only be achieved with an extensive distributed reservation which entails an enormous

---
[1]The sample size k in this context should not be confused with k in the context of a $k-$ *neighborhood*.

8.3. PRECISION AND RECALL IN RANDOM NETWORKS

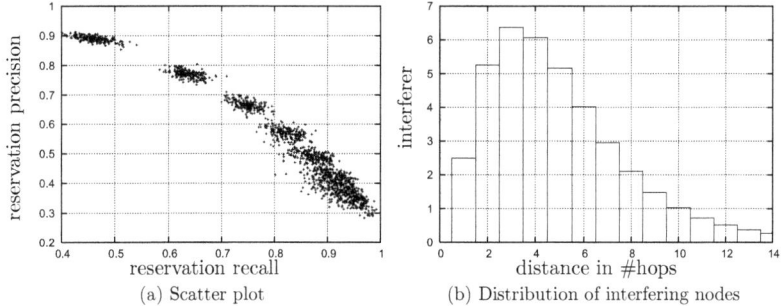

(a) Scatter plot   (b) Distribution of interfering nodes

Figure 8.3: Recall and precision trade-off

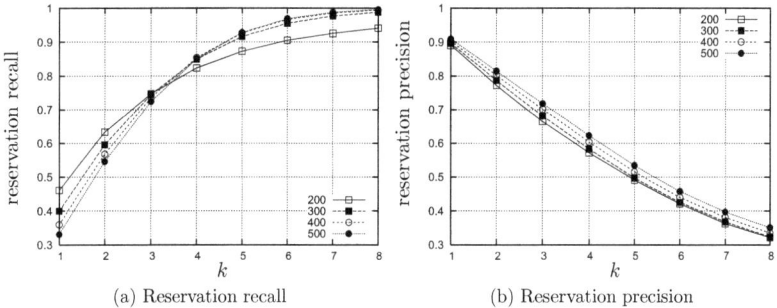

(a) Reservation recall   (b) Reservation precision

Figure 8.4: Impact of network density

waste of bandwidth. Why it is unfeasible to choose a value for k that maximizes both reservation recall and reservation precision is shown in Figure 8.3b. The Figure shows the average distribution of the interfering nodes with respect to their hop distance measured from the node they interfere with. As can be seen, the distribution's peak is around 3 hops. Choosing a value of 3 for k, however, does not take the tail of the distribution into account. Since the tail is not negligible, a large set of interferer is not covered. Choosing a value of 10 or 11 covers the nodes at the tail of the distribution, but at the same time, includes many nodes which are not interfering at all.

8.3.3 Impact of network density

We now explore how recall and precision are affected by the network density. We look at a network of size $2000m \times 2000m$ while deploying an increasing number of nodes. As can be inferred from Figure 8.4a, reservation recall drops with an increasing network density if k is smaller than 3, but increases with the network density if k is greater than 3. This behavior is a direct consequence of the fact that the number of nodes in a disc grows with the square of the radius of the disc. Imagine a simplified view where the k-neighborhood is represented by all nodes located in a disc d_k with

CHAPTER 8. DISTRIBUTED BANDWIDTH RESERVATION

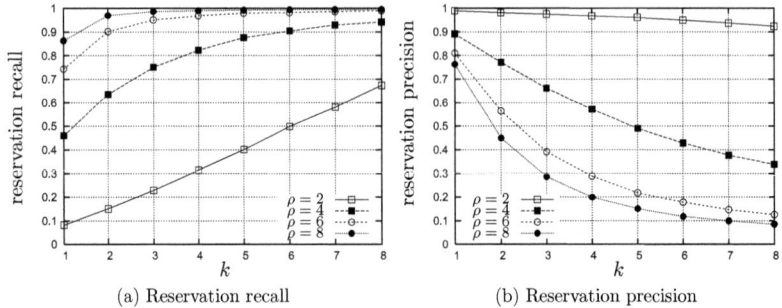

(a) Reservation recall   (b) Reservation precision

Figure 8.5: Impact of pathloss coefficient

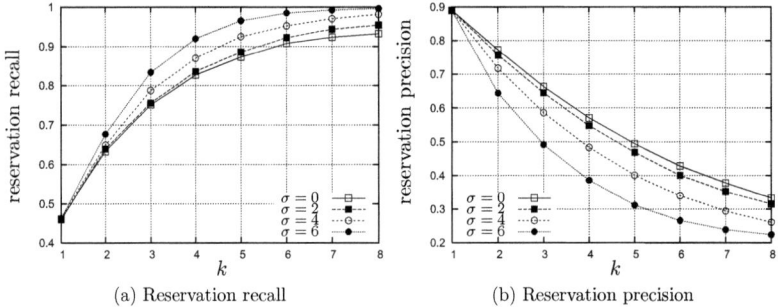

(a) Reservation recall   (b) Reservation precision

Figure 8.6: Impact of shadowing

radius r_k and all interfering nodes to be the nodes located in another disc d_I with radius r_I. If the network density grows, obviously the number of nodes within d_I grow faster than the number of nodes in d_k, if $r_I > r_k$. This is what happens in Figure 8.4a when k is smaller or equal than 3. Since with increasing node density the number of interfering nodes increases faster than the number of nodes within the set k-NH, the reservation recall drops. If, on the other hand, k is greater than 3, then r_k becomes larger than r_I and the opposite happens: with increasing node density, the number of nodes in k-NH increases faster than the number of interfering nodes, which improves the reservation recall. While the influence of the network density on reservation recall very much depends on the value of k, its effect on reservation precision is more consistent and also more modest. From Figure 8.4b we see that the reservation precision slightly improves as the network density increases. The reason is that the set of interfering nodes grows a little faster than the k-neighborhood for an increasing network density.

8.3.4 Impact of radio propagation

So far, we fixed the path loss and the shadowing deviation of the radio propagation (Equation 3.1) to be 4 and 0 respectively. However, we have seen in the previous chapter that both path loss and shadowing deviation have an impact on the interference perceived at a given node. Figure 8.5 illustrate the effect of the path loss coefficient on recall and precision. It is shown that the bigger the path loss coefficient, the higher the recall (at the expense of a lower precision). The reason is that a high path loss coefficient makes the signal drop below the interference threshold very quickly and thus makes the set of interfering nodes approach the 1-hop neighborhood. That's why – already for a k of 1 – the recall for a path loss coefficient of 8 becomes bigger than 0.8. On the other hand, the precision drops very quickly for high path loss values (Figure 8.5b). This is because most nodes outside the 1-hop neighborhood no longer.

Figure 8.6a shows how reservation recall evolves when the radio propagation becomes irregular. From the Figure, we observe that for a given value of k, the reservation recall improves as σ increases. The reason is, as mentioned in chapters 4 and 7, the increasing σ in the LNS model also increases the transmission range of the node. It follows that the set of decoders $\mathcal{D}_n$ increases and so does the set of of k-neighbors, which improves the overall reservation recall. How the reservation precision is affected by irregular radio propagation is illustrated in Figure 8.6b. It can be observed that the reservation precision drops as σ increases. This is what we would expect since the higher the values for σ the more randomness is induced into the signal. As the randomness increases, the correlation between the number of hops of a source destination pair and its Euclidean distance is reduced.

8.4 End-to-end reservations

Reservation recall and precision give an indication of the quality of a reservation for one single transmission. In practice, one is interested in the quality of a reservation of a whole flow.
Bandwidth and throughput capacity are concave metrics. The throughput of a flow is determined by the lowest bandwidth available along the flow. Similarly, the quality of a flow is determined by the lowest reservation recall along the path. We therefore define a new metric, the so called *reservation quality* $Q^k_{n',n}$ for a flow with node n' being the source node and node n being the destination node:

$$Q^k_{n',n} = \min_{(n''',n'')\in\eta(n',n)} R^k_{n''\leftarrow n'''} \qquad (8.5)$$

As with reservation recall and precision for a single transmission, we also want to consider the downside of a high quality reservation, namely the amount of resources wasted along the path from node n to node n'. This is reflected by the so called *reservation loss* $L^k_{n',n}$

$$L^k_{n',n} = \frac{\sum_{(n''',n'')\in\eta(n',n)} \mid \text{k-NH}_{n''} \cap \mathcal{I}_{n''\leftarrow n'''} \mid}{\sum_{(n''',n'')\in\eta(n',n)} \mid \text{k-NH}_{n''} \mid}. \qquad (8.6)$$

Obviously, if $\mid \eta(n',n) \mid = 1$, both the reservation quality and reservation loss are

66   CHAPTER 8. DISTRIBUTED BANDWIDTH RESERVATION

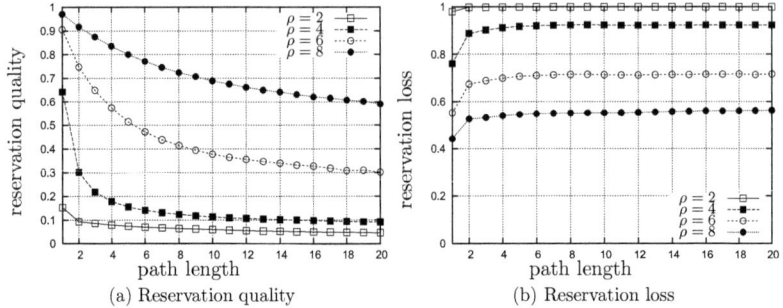

Figure 8.7: Reservation quality and loss vs path length

identical to reservation recall and precision respectively. As with capacity (Equation 5.13), reservation quality and loss are defined over a traffic pattern Υ (Equation 5.9). Thus, for a given set of flows Υ, $\frac{1}{k \cdot |\Upsilon|} \sum_{i=0}^{k-1} \sum_{(n',n) \in \Upsilon} Q_{n',n}^k |_{X=x^i}$ is an approximation for the expected reservation quality of a network. Similarly, $\frac{1}{k \cdot |\Upsilon|} \sum_{i=0}^{k-1} \sum_{(n',n) \in \Upsilon} L_{n',n}^k |_{X=x^i}$ is an approximation for the expected loss of a network.

8.5 Reservation quality and Loss in random networks

We consider uniform random network topologies of 200 nodes being deployed in an area of $2000m \times 2000m$ and we use the very same settings for interference, transmission power, routing and scope as in section 8.3. Now, we are interested in how the reservation quality Q^k and loss L^k behave as the length of the path increases.

8.5.1 Trade-off between reservation quality and loss

Figure 8.7 shows both Q^k and loss L^k as a function of the path hop length $| \eta(n,n') |$ for $k = 2$ and for different pathloss values ρ. A first observation is that, for a reasonably long path, the reservation quality is beyond repair. The result tells us that there is a high probability that a reservation along the path covers only about 20% of all interfering nodes. As expected, increasing ρ improves the reservation quality at the expense of the reservation loss. Similar to reservation recall and precision, there is a trade-off between reservation quality Q^k and reservation loss L^k (Figure 8.7b). The Figure also shows that if the path length increases, the curves start to resemble a linear function. This implies that if the path is long enough, any change in ρ equally affects Q^k and L^k. There is one notable distinction in the behavior of reservation quality Q^k and loss L^k in Figure 8.7. While reservation loss converges very quickly, the reservation quality keeps dropping with increasing path length, particularly for a ρ of 8 or 6. This is due to the concave property of the Q^k metric which defines the reservation quality as the minimum reservation recall along the path.

8.5. RESERVATION QUALITY AND LOSS IN RANDOM NETWORKS 67

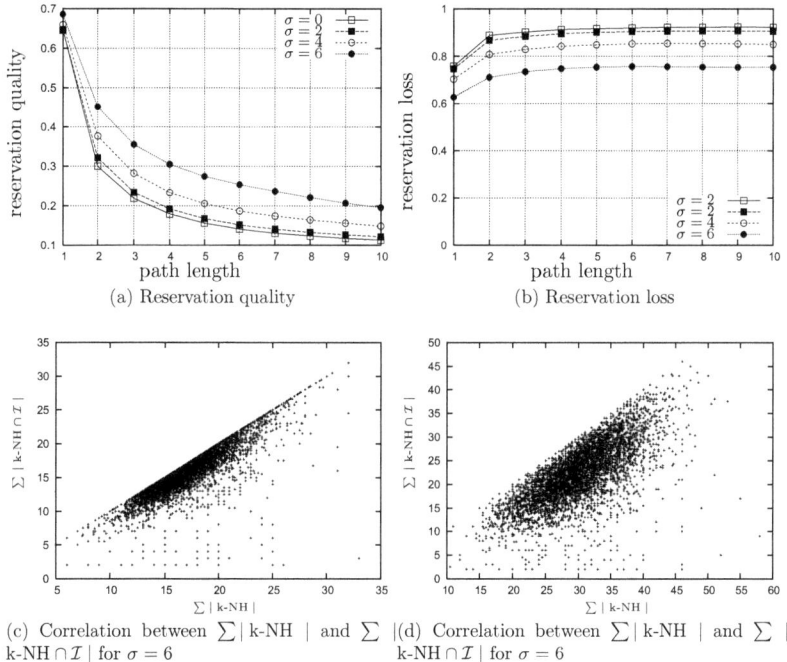

(c) Correlation between $\sum|$ k-NH $|$ and $\sum|$ k-NH $\cap \mathcal{I}|$ for $\sigma = 6$

(d) Correlation between $\sum|$ k-NH $|$ and $\sum|$ k-NH $\cap \mathcal{I}|$ for $\sigma = 6$

Figure 8.8: Reservation quality and loss

8.5.2 Effects of fading

Figure 8.8 illustrates the effect of fading on reservation quality and loss for $k = 2$. In section 8.3 we mentioned that shadowing increases the average signal power per distance and therefore augments the set of direct neighbors $\mathcal{D}_n$ of a node n. This has an augmenting effect on reservation recall and, as a consequence, also increases the reservation quality of a flow (Figure 8.8a). The reservation loss L on the other hand, decreases with increasing values of σ, similar to reservation precision (Figure 8.8b). A look at the definition of reservation loss (Equation 8.6) explains what happens. Reservation loss relates the sum of all interfering k-neighbors to the sum of k-neighbors along the path. Figures 8.8c-d show the correlation between these two components for $k = 2$ and different values of σ. No loss at all would require the sum of interfering k-neighbors to match the sum of k-neighbors. In such a case, the correlation plot would look like a straight line with a slope of one. From Figures 8.8c-d we see that with increasing values of σ, the correlation plot progressively shifts from a line-like shape to a cloud. The cloud in Figure 8.8d emerges if there is no real correlation between the sum of interfering k-neighbors and the sum of k-neighbors, which obviously is the case as σ grows.

8.6 Summary

In this chapter we studied the bandwidth reservation in ad hoc networks. To do so, we have defined to two metrics of interest: Reservation recall which models the quality of a reservation. And reservation precision which models the amount of resources that are wasted due to a reservation. We have shown that there exists a clear trade-off between reservation recall and precision since optimizing recall is done at the cost of precision. This is because there is no exact correlation between the k-neighborhood and the set of interferer for a given transmission. A large k-neighborhood increases the interference coverage but induces over-reservation. In the chapter, we have also shown that irregular radio propagation diminishes the reservation precision but improves the recall. One reason for this is that the set of nodes in a k-neighborhood grows under fading, which then increases the interference coverage. Recall and precision have a direct impact on the quality of a bandwidth reservation along the path of a flow. We have shown that even for a 3-hop wide reservation, there is a high probability that a reservation along the path covers only about 20% of all interfering nodes. In this work, we considered only static networks. One could imagine that the situation deteriorates even more in mobile scenarios, where nodes occupying a reservation leave the interfering area at some point in time. The conclusion is that distributed bandwidth reservations are inadequate to provide QoS in MANETs and that other techniques have to be considered, e.g., priority based approaches.

Part II
System

Chapter 9
The invisible ad hoc network

After having discussed some of the more formal and theoretical problems of ad hoc networks in part I of this dissertation, we want in this part to look at ad hoc networks from a systems perspective. In chapter 1 we have described some of the possible application scenarios of ad hoc networks. Unfortunately, up to now, only a few real deployments of ad hoc networks exist, and most of them are at Universities or research laboratories. One reason for this is that there are not enough suitable applications available for ad hoc networks. In order for certain applications to run in MANETs they have to be able to cope with the complexity of the underlying ad hoc network, be it the error prone wireless media or the lack of infrastructure and central components. In an effort to overcome the absence of practical applications for MANETs, researchers have designed and implemented several dedicated applications that were tailored towards the requirements of MANETs. Those applications particularly take the very specific properties of MANETs into account, but at the same time, their usage is restricted to MANETs only, which prohibits access to a broader audience. In practice, users prefer to use the same application regardless of the network they are currently part of. We believe that, rather than building dedicated applications, one should provide a network infrastructure to allow existing applications to run seamlessly in ad hoc networks. In order to implement this goal, we present in this part of the dissertation, network abstractions and services that hide the complexity of the underlying ad hoc network to higher software layers. Our approach can be seen as an effort to turn MANETs into an *invisible network* to open access for traditional Internet based applications.

In the first chapter of this part of the thesis we present the design and implementation of a virtual interface that allows nodes running different media access technologies to form one single ad hoc network. Thereby, the virtual interface is hiding all the heterogeneity of the different wireless media access technologies below a standard IP interface. We evaluate our approach in terms of throughput and handover-time using IEEE 802.11 and Bluetooth as two examples of wireless access technologies.

In the remainder of this thesis, we present the architecture of several fundamental network services for MANETs. Many distributed applications that potentially would suit for MANETs expect certain services like DNS or SIP to run. However, the inherent distributed and infrastructure-less nature of MANETs has highlighted how fundamental services, such as DNS, SIP, SLP, etc., rely on a centralized client-server model. The absence of those basic services in MANETs does not strictly prevent networking, but it strongly restraints the adoption of ad hoc networks as a plug-and-play technology. Ideally, traditional Internet-based applications would run seamlessly in ad hoc networks, without the need for any change re-configuration. One approach to achieve this goal is to provide distributed versions of network services in MANETs, while still maintaining the standardized interface of those services.

Recently, there have been many different attempts at implementing network services in MANETs using a variety of technologies including, e.g., extensions to routing protocols [75] and distributed hash tables (DHTs) [76, 77, 78, 79]. In practice, most of these solutions are either not efficient enough or they introduce severe limitations on how the MANET can be operated. For instance, some solutions only work on a particular routing protocol [75], other solutions require a fixed node to be stable and available at all times [76, 79], and yet a third set of solutions treat MANETS like the Internet and try to build overlay networks that end up being too expensive to maintain [78, 79].

In this thesis, we take a radically different approach to the problem. We first focus on the fundamental infrastructure needed to implement a basic directory service on MANETs. Briefly explained, sharing information in MANET can be seen as the problem of dynamically storing and performing lookups over key/value pairs. We have developed and implemented a system called *MAND* which provides mechanisms to store and lookup tuples (key/value pairs) over a MANET. *MAND* exploits the fact that MANET routing protocols periodically exchange information using a variety of routing messages. Given that finding a service or a particular piece of information in a MANET requires to establish a route to wherever the service is located, *MAND* piggybacks the message traffic required to spread information across the MANET on the routing messages, thereby requiring no additional traffic. At the same time, *MAND* does not impose any constraints on the type of routing protocol that is used in the network, nor does it require the routing protocol to be modified or re-compiled.

We use the *MAND* directory service to build more complex services with broader semantics. In chapter 12 we present a fully decentralized and SLP compatible service discovery infrastructure for MANETs. We also show how such a system can be used to implement an efficient MANET-Internet gateway solution. In chapter 13, we present *DOPS* (Domain Name and Presence Service for Ad Hoc Networks), a middleware platform for domain name resolution in MANETs which also provides seamless interaction with the Internet. In chapter 14, we describe SIPHoc, a middleware platform for session establishment and management in MANETs. To demonstrate the potential of SIPHoc and its feasibility, we have used SIPHoc to provide a VoIP solution that supports VoIP conversations within a MANET and between the MANET and end-points on the Internet. In chapter 16 we present an application for social networking in ad hoc networks.

The design and implementation of all the services presented in this thesis is tailored to MANETs of a reasonable size of up to 100 nodes at maximum. The foundation of this assumption lies in part I of this thesis where we have shown how throughput capacity scales with the network size. We believe that throughput capacity is the major limiting factor that dermines to what size ad hoc network will scale.

To emphasize the practical relevance of our work, we have implemented all the services proposed in this thesis in Linux. Our deployment testbed consists of 23 Linux driven Nokia N800 handheld devices. Experiments have been done to study the reliability and the performance of the systems.

Chapter 10

Heterogeneous Mobile Ad Hoc Networks

Ad Hoc Networks are considered to be formed spontaneously between various different wireless devices. The devices participating an ad hoc network may vary in terms of CPU performance, memory capacity or bandwidth. In some cases, the devices may even be equipped with different media access technologies. Current media access proposals for wireless networks range from Bluetooth [80] and IEEE 802.15.4 [81] up to 802.11 [82]. Already within the 802.11 group there exist many different flavors (802.11e/g/b/a/h). Nevertheless, most research on MANETs assumes such networks to have one common MAC schema. In this chapter, we discuss the integration of heterogeneous mobile ad hoc networks comprising different MAC protocols. We describe and discuss the performance of an end-to-end communication abstraction that transparently hides the heterogeneity of the underlying medium access schema while still supporting node mobility, multi-hop transmission, simple addressing, etc. Such a heterogeneous mobile ad hoc networks occurs, e.g., when combining a Bluetooth PAN with an 802.11 MANET as shown in Figure 10.1. An example is a university campus with all sorts of personal devices ranging from mobile phones, handhelds up to laptops. Each of these devices may be equipped with different communication technologies tailored to their capabilities and intended use. Bluetooth, e.g., is an energy-saving technology commonly used in mobile phones or handhelds. Laptops on the other hand do not have such strong constraints, they might include an 802.11 as well as a Bluetooth interface. Ubiquitously combining all these devices into one heterogeneous mobile ad hoc network could invite new application and services like, e.g., location based services or VoIP. In such scenarios, a personal device of one particular PAN might communicate with a personal device of another PAN in a multi-hop fashion where the underlying MAC scheme changes on a per hop basis. Another interesting application of a heterogeneous mobile ad hoc network is the integration of various personal devices into a so called digital home. In a digital home, PCs and consumer electronics could work together on an ad hoc basis to different rooms in the house using a combination of networks.

The rest of this chapter is organized as follows. The next section explores various approaches to build heterogeneous mobile ad hoc networks using Bluetooth and 802.11. Section 10.3 presents our solution which is based on virtual layer two device. In section 10.4 we evaluate the system and section 10.7 summarizes the chapter.

CHAPTER 10. HETEROGENEOUS MOBILE AD HOC NETWORKS

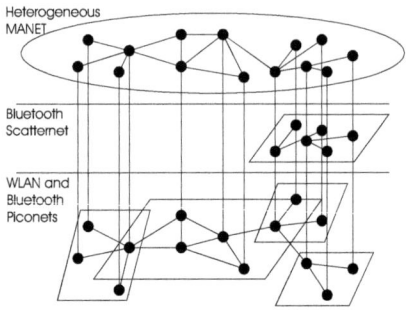

Figure 10.1: Heterogeneous Mobile Ad Hoc Network

10.1 PolyMAC in MANETs using Bluetooth and 802.11

As mentioned we aim at the integration of heterogeneous mobile ad hoc networks comprising different MAC protocols. Although we look for a generic solution we take a network as shown in Figure 10.1, including 802.11 and Bluetooth, as a base for our investigations.

802.11 is currently the most common MAC protocol used in MANETs as it has achievable data rates of up to $54 Mbits/s$. From the two modes of operation only the decentralized *contention based* mode is suitable for MANETs. Medium access is then provided by a CSMA/CA protocol. Bluetooth on the other hand is optimized for low cost and only offers data rates of $\sim 1 Mbit/s$. Media Access in Bluetooth is centrally organized. A master node manages the time slots within a *piconet* (collaboration between maximum 8 nodes), thereby avoiding collisions. Several piconets can be combined to a so called *scatternet*. In a scatternet packets no longer follow a predefined way towards their destination. Instead they have to be routed on a per master basis. From a more abstract perspective we could also say, a scatternet is a MANET where the piconets are the nodes. Besides these characteristics Bluetooth and 802.11 differ with regard to message broadcast. While 802.11 inherently supports broadcast, Bluetooth does not support it at all [1].

After having described the specific properties of 802.11 and Bluetooth we now want to define the requirements which needs to be fulfilled by a heterogeneous mobile ad hoc network. Basically we think such a network should at least satisfy the following criteria:

1. **Transparency:** The network should provide transparent end-to-end communication.

2. **Mobility:** Node mobility has to be supported.

---
[1] Bluetooth 2.0 specifies an Active Slave Broadcast (ASB). But up to our knowledge there is no implementation supporting it.

10.2. OVERVIEW

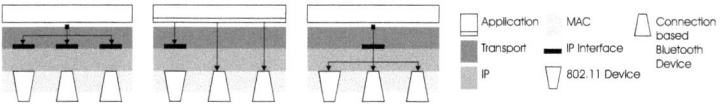

Figure 10.2: Communication Abstraction: (a) Transport Layer (b) Application Layer (c) IP Layer

3. **Addressing:** Addressing should be independent of how many interfaces are attached to a particular node.

4. **Configuration:** Configuring a node should be possible without any knowledge of the underlying MAC technology.

In a nutshell, what we are looking for is an end-to-end communication abstraction that supports MAC-switching[2], node mobility and multi-homing [3]. Certainly two issues to be solved are broadcast emulation and handover. Broadcast emulation since it is not directly supported in Bluetooth (nor on nodes comprising both Bluetooth and 802.11). Handover since in the case of heterogeneous mobile ad hoc networks a handover might include a change in how the medium is accessed. A handover can be caused by node mobility, a change in user preferences (the user chooses to save energy and use Bluetooth instead of 802.11), or performance reasons. We will come to that point later in section 10.3.

10.2 Overview

In principle, a communication abstraction like the one proposed in the previous section can be provided on any layer above MAC. SCTP [83], or Stream Control Transmission Protocol, is a transport protocol defined by the IETF providing similar services to TCP. It ensures reliable, in-sequence transport of messages. While TCP is byte-oriented, SCTP deals with framed messages. A major contribution of SCTP is its multi-homing support. One (or both) endpoints of a connection can consist of more than one IP addresses, enabling transparent fail-over between hosts or network cards. Since both, 802.11 and Bluetooth have IP bindings, one could think of using a common MANET routing protocol. Each interface could be separately configured and maintained (AODV-UU [84], e.g., supports multiple interfaces). This solution (see Figure 10.2a) seems to be quite promising in terms of performance since SCTP optimizes the transmission over multiple links. In fact, if one particular node can be reached through several interfaces, SCTP switches transmission from one interface to another after a predefined number of missing acknowledgements. Unfortunately, the solution lacks transparency. Applications running traditional Unix sockets would have to be changed to use SCTP sockets instead. Another problem arises with the connection oriented nature of Bluetooth. In Bluetooth, interfaces appear and disappear dynamically depending on whether the connection to the specific node is currently up or down. Therefore, this is something that both the ad hoc routing

---

[2]Refers to the fact that the used MAC technology may change along a source/destination path
[3]A node having multiple network interfaces

protocol as well as SCTP would have to cope with. Recently, the Network Working Group at the IETF proposed ASCONF [85], a feature that allows dynamic reconfiguration of IP addresses during an SCTP session. On the routing layer, however, we are not aware of any protocol that supports dynamic interfaces. Typically, these protocols expect the network interfaces to be configured statically within the system. And even if routing modules were able to handle dynamic interfaces, still the mapping between the nodes and IP addresses would have to be solved. Instead of using an all-IP communication based on SCTP and a MANET routing protocol, one could as well think of an application layer solution (see Figure 10.2b). In [86], the authors propose a communication framework based on JXTA [87, 88] that is in charge of transporting packets end-to-end by choosing the right transport mechanism below, i.e 802.11/IP or raw Bluetooth. This allows the use of separate optimized routing algorithms for IP and Bluetooth respectively. Unfortunately, this solution goes at the expense of performance. How should the application layer framework decide which subsystem best to send the packet to without using flooding? Furthermore, application transparency is not given at all. Rather, transparency can be provided when multi homing is implemented on a lower layer. Imagine a single IP interface (virtual interface) handling different physical interfaces (see Figure 10.2c). Neither the application nor the routing would be affected by the fact that eventually a node consists of multiple interfaces belonging to different MAC technologies. But how can such an interface know about the real physical interface to send the packet to? How feasible is such a solution? The following sections not only shows that such a solution is feasible, but also that it performs well.

10.3 A MAC Layer Approach

As discussed, a *virtual interface* approach is the most convenient solution for heterogeneous mobile ad hoc networks in terms of transparency. The question now is how to implement it and how to make it perform well.

10.3.1 Linux Ethernet Bridge

The solution we propose here is inspired by the Linux Ethernet bridge [89, 90]. Similar to a physical bridge device, the Linux Ethernet bridge ties separate layer-two-networks together, only that it is purely software. The bridge includes much of the functionality we would like to see in a *virtual interface* as described at the end of section 10.1. Appearing to the operating system as a regular layer-two-device one can easily assign IP addresses to bridges. The bridge is supposed to work in combination with 802.x devices, therefore not including Bluetooth. Fortunately, the Bluetooth personal area network (PAN) profile specifies BNEP [91] which itself defines a packet format to transport common networking protocols over the Bluetooth media. BNEP supports the same networking protocols that are supported by IEEE 802.3/Ethernet encapsulation, therefore enabling Bluetooth to be used within the bridge. Layer-two-bridges have the well known property of forwarding packets to the correct physical network corresponding to the destination MAC address. So the bridge's decision for a frame is one of these:

- bridge it, if the destination MAC address is on another side of the bridge

10.3. A MAC LAYER APPROACH

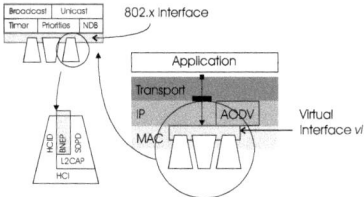

Figure 10.3: Virtual Interface Approach

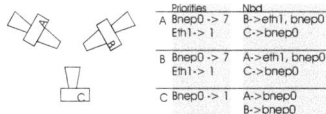

Figure 10.4: The Neighborhood Database (NDB)

- flood it over all the forwarding bridge ports, if the position of the box with the destination MAC is unknown to the bridge

- pass it to the higher protocol code (IP) if the destination MAC address is that of the bridge or of one of its ports

- ignore it if the destination MAC address is located on the same side of the bridge

As the bridge learns about the neighboring nodes and their MAC addresses, it is capable of forwarding the packet to the right physical interface instead of just flooding the packet. It does so by keeping a neighborhood database (NDB) mapping destination MAC addresses to physical interfaces. Entries are refreshed by incoming packets, otherwise they time out.

10.3.2 The virtual Interface

Since we are looking for a virtual interface, we are not interested in packet forwarding, but the idea of storing a MAC/interface mapping based on incoming packets is also suitable for local traffic. We have implemented a virtual interface (vi) that adopts this mechanism. Like the Linux Ethernet bridge, the vi represents a regular layer-two-device and can be configured accordingly. The vi allows to plug in any 802.x compatible network device, like e.g a wireless LAN card or a BNEP/Bluetooth connection, while hiding the heterogeneity of the used devices from the upper layers. For every neighboring node, the vi holds an array of possible outgoing interfaces. The number of entries in the array follows directly from the intersection of the two sets of interfaces. An entry contains a time-stamp and is created upon receiving the first packet (i.e a routing broadcast message or a route reply) of the associated neighbor/interface pair. Every consecutive incoming packet refreshes the time-stamp.

Algorithm 10.1 Local packet processing

1: receiving_packet(vi, p);
2: **if** p.type = BROADCAST **then**
3:    interfaces := get_interfaces(vi);
4:    **for all** i in interfaces **do**
5:       transmit(i, p);
6:    **end for**
7: **else**
8:    current := ndb[hashval(p.mac.dst_address)];
9:    outgoing := NULL;
10:    **while** current $\neq$ NULL **do**
11:       **if** outgoing = NULL **then**
12:          outgoing := current;
13:       **else if** current.priority < outgoing.priority **then**
14:          Δ := outgoing.ts - current.ts;
15:          **if** Δ < vi.maxdiff **then**
16:             outgoing := current;
17:          **end if**
18:       **else**
19:          Δ := current.ts - outgoing.ts;
20:          **if** Δ > vi.maxdiff **then**
21:             outgoing := current;
22:          **end if**
23:       **end if**
24:       current := current.next;
25:    **end while**
26:    **if** outgoing $\neq$ NULL **then**
27:       transmit(outgoing, p);
28:    **end if**
29: **end if**

The processing of outgoing packets is done using Algorithm 10.1. If the vi receives a packet from the upper layer for delivery, it first checks the packet type (line 2, Alg. 10.1). In case the packet is a broadcast packet, it will be sent through all available interfaces (lines 4-6, Alg. 10.1). Therefore, the vi also acts as a broadcast emulation layer for Bluetooth. This is particularly important since BNEP is point-to-point. However, if the packet is unicast (lines 7-29, Alg. 10.1), the vi looks for the corresponding entry in the neighborhood database (line 8) mentioned above and retrieves the information about the interface the packet has to be sent to (entries are periodically checked for expiration). If there is more than one option, the vi makes use of another feature, the so called priority table. The priority table specifies some sort of ranking among the interfaces, meaning that whenever a given neighbor can be reached through several interfaces, the interface with the lowest priority is taken. Figure 10.3 illustrates the architecture of the virtual interface and how it is embedded within the network stack. The state of the neighborhood database and priority table in a scenario where three nodes form a neighborhood (they are within transmission range of each other) is shown in Figure 10.4.

Apart from the priority table, another mechanism is needed to prevent links from becoming stalled. Imagine two nodes, each of them containing an 802.11 as well as a Bluetooth interface. Obviously, there are two ways for the two nodes to exchange packets, therefore the corresponding neighborhood database contains two entries. Further, assume the Bluetooth interface to have the highest priority on the sender node (the node might be battery driven and tries to minimize energy consumption per packet transmission). If the Bluetooth connection temporarily suffers from very bad link properties, the node probably wants to switch to the 802.11 interface at the expense of a higher energy consumption. This will not be possible under the current setup as the *vi* always uses the interface with the highest priority to transmit the packet. To overcome this unfavorable situation, a cleanup timer, which periodically removes all expired entries, could be used (an entry is expired if it has not been refreshed for a certain amount of time), but this would unnecessarily increase routing traffic since the next packet in the queue right after the cleanup is triggering a route request. Instead of a cleanup timer, we therefore use a different mechanism that is sort of routing-friendly. We introduce a new parameter that is associated with a *vi*, the so called *maxdiff* threshold. The *maxdiff* threshold unit is 10ms and it decides how much two single entries within the neighborhood database may differ in terms of time-stamps to keep the priority policy up. So a higher ranked interface entry can be replaced by a lower priority interface if the time-stamp differs for more than *maxdiff* (line 15, line 20, Alg. 10.1).

10.3.3 Configuration

The virtual interface consists of a kernel module and a user-mode command-line tool [4]. The kernel module immediately gets loaded when creating a new virtual interface using the *victl addvi* command. There are also commands to add and remove devices to/from the virtual interface as well as to set the interface priorities. Table 10.1 shows how a virtual interface containing Bluetooth and WLAN is typically set up. Like a regular layer-two-interface, the IP configuration of vi can be performed with *ifconfig*. The question that remains is how to dynamically add Bluetooth devices to the *vi* (remember that they appear on a connection basis). Apparently, we need some help from Bluez [92] here. A script is called whenever a Bluetooth interface is created. We then simply let the *vi* maintenance be triggered by this event, as shown in Figure 10.2. Devices are added and removed to/from the *vi* every time they appear or disappear. Another issue to take care of is the MAC address of the *vi*. The MAC address can manually be set using the *victl addr* command. This address can be totally virtual or correspond to a physical interface. Outgoing packets always adopt the MAC address of the *vi*, regardless of the interface they leave. Some wireless NICs do not allow MAC addresses to be overwritten, so the address for the *vi* is best chosen to correspond to the address of NIC that cannot be changed. On the contrary in a *vi* comprising Bluetooth interfaces only, a virtual MAC address is more suitable since otherwise, communicating nodes would have to adapt their ARP binding every time the MAC address changes.

10.3.4 Handover

A handover includes route changes as well as MAC switching. In principle, there are three possible scenarios (see also Figure 10.5):

---

[4]Furthermore a C API is provided for the case where the application directly wants interact with the *vi*.

CHAPTER 10. HETEROGENEOUS MOBILE AD HOC NETWORKS

```
someuser@somedesk:~$ victl help
commands:
    help                                    command list
    addvi     <vi>                          add vi
    delvi     <vi>                          delete vi
    addif     <vi> <device>                 add interface
    delif     <vi> <device>                 del interface
    setdiff   <vi> <maxdiff>                set maxdiff
    setprio   <vi> <port> <prio>            set priority
    addr      <vi> <mac-addr>               set mac address
    show      <vi>                          list of vi's
someuser@somedesk:~$ victl add vi0
someuser@somedesk:~$ victl addif vi0 eth1
someuser@somedesk:~$ victl setprio vi0 eth1 80
someuser@somedesk:~$ victl addif bnep0
someuser@somedesk:~$ victl setprio vi0 bnep0 100
someuser@somedesk:~$ ifconfig eth1 0.0.0.0
someuser@somedesk:~$ ifconfig vi0 192.168.220.2
someuser@somedesk:~$ victl addr vi0 00:02:72:B2:78:DC
someuser@somedesk:~$ victl setdiff 200
someuser@somedesk:~$ ifconfig
bnep0  Link encap:Ethernet  HWaddr 00:02:72:B2:78:D2
       UP BROADCAST RUNNING MULTICAST  MTU:1500  METRIC:1
       RX packets:0 errors:0 dropped:0 overruns:0 frame:0
       TX packets:4 errors:0 dropped:0 overruns:0
       collision:0 txqueuelen:100
       RX bytes:104 (104.0 b) TX bytes:88 (88.0 b)

eth1   Link encap:Ethernet  HWaddr 00:02:2D:7B:88:D1
       UP BROADCAST RUNNING MULTICAST  MTU:1500  METRIC:1
       RX packets:1093 errors:277 dropped:0 overruns:0
       TX packets:51 errors:0 dropped:0 overruns:0
       collision:0 txqueuelen:100
       RX bytes:65778 \{64.2 Kb\} TX bytes:11386 \{11.1 Kb\}
       Interrupt:11 Base address:0x100

vi0    Link encap:Ethernet  HWaddr 00:02:72:B2:78:DC
       inet addr:192.168.220.2 Bcast:192.168.220.255
       UP BROADCAST RUNNING MULTICAST  MTU:1500  METRIC:1
       RX packets:0 errors:0 dropped:0 overruns:0
       TX packets:0 errors:0 dropped:0 overruns:0
       collisions:0 txqueuelen:0
       RX bytes:0 (0.0 b) TX bytes:0 (0.0 b)

someuser@somedesk:~$ victl show vi0
virtual interface maxdiff interfaces  priority
vi0               200     bnep0       100
                          eth1        80
someuser@somedesk:~$
```

Table 10.1: Setting up a virtual interface

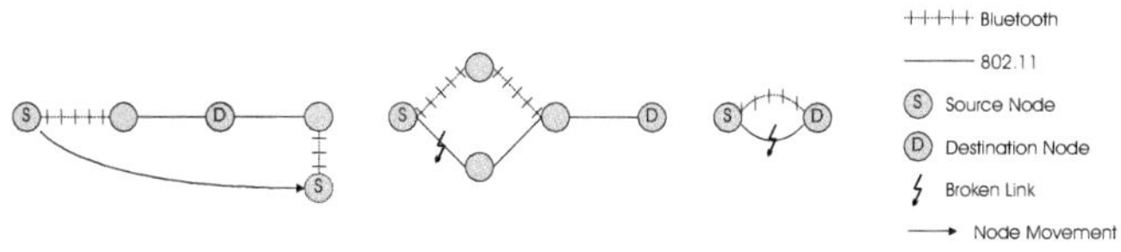

10.3. A MAC LAYER APPROACH

Figure 10.5: Horizontal (1), Diagonal (2), and Vertical (3) handover

```
#!/bin/sh
victl addif vi0 \$1
victl setportprio vi0 \$1 80
ifconfig \$1 0.0.0.0
```

Table 10.2: /etc/bluetooth/pan/dev-up

1. **Horizontal Handover:** the route changes while the underlying MAC technology remains the same

2. **Vertical Handover:** the route does not change but the given neighbor is now reached through a new physical interface

3. **Diagonal Handover:** MAC technology and route change simultaneously

Let us now take a closer look at how routes get established in a heterogeneous mobile ad hoc network when using virtual interfaces. In the case of a reactive routing protocol (e.g. AODV [93]), it is the application that triggers a path setup. Since there is no route available yet, the routing protocol typically first broadcasts a route request. At the very beginning the neighborhood database contains no entries but the transmission of a broadcast packet does not need any neighborhood information anyway (see section 10.3). After the route request has passed several hops, a route reply eventually returns back to the origin. The route reply not only establishes the route but also creates an entry within the neighborhood database, providing the vi with information on the interface to which the packets to the given neighbor have to be transmitted. In the case of a pro-active routing protocol (e.g. OLSR [94]), things are slightly different. Here nodes periodically broadcast their neighboring information and therefore are also creating entries within neighborhood databases. In both cases - proactive and reactive - the NDB entry is established in combination with the new route, regardless of whether the MAC technology changes or not. In a vertical handover on the other hand, the NDB is affected differently. Either the priority table changes due to user preferences or certain entries expire, meaning their timestamp difference exceeds the threshold $maxdiff$ (bad link quality). Obviously, the smaller $maxdiff$, the more agile the virtual interface is for handover. But a small $maxdiff$ threshold can also be unfavorable since it is a counter movement to the priority policy. If the priorities reflect the bandwidth of the interfaces, then a small $maxdiff$ threshold may lead to a diminishing throughput, as we will see in section 10.4.

Type	From	To	Routing	Mode	Avg. Packet Lost	Stddev
Horizontal	WL	WL	AODV	raw	2	0
				vi	2	0
			OLSR	raw	2.1	0.7416
				vi	2.1250	0.8539
	BT	BT	AODV	vi	34.6	2.67
			OLSR	vi	44.3	8.06
Diagonal [5]	WL	BT	AODV	vi	2.2	0.42
			OLSR	vi	4.5	1.28
Vertical	WL	BT	AODV	priority driven	< 1	0
				vi/1	0.6	0.2108
				vi/200	1.4	0.3944
				vi/500	4.3	0.2581
				vi/1000	9.6	0.2108
			OLSR	priority driven	< 1	0
				vi/1	< 1	0
				vi/200	2.2	0.5374
				vi/500	4.3	0.7888
				vi/1000	10.25	0.6770
			—	priority driven	–	–
				vi/1	20.1	0.316
				vi/200	20.2	0.421
				vi/500	20	0
				vi/1000	20.1	0.316

Table 10.3: Handover-time

10.4 System Evaluation

This section describes some of the experiments in a heterogeneous environment based on the *virtual interface*. Our setup consists of up to five DELL Latitude laptops. We used PCMCIA Cisco aironet cards with 802.11b running and Acer Bluetooth dongles implementing the Bluetooth 1.1 specification. The 802.11 NICs were switched to *ad-hoc* mode. All the laptops were located in one room without any other wireless devices turned on simultaneously. To create different network topologies an artificial setup using packet filter rules was used: the laptops were only allowed to communicate with their direct neighbors , all other traffic was dropped (by the default policy). We have used the Uppsala implementation of AODV [84] and the OLSR implementation in [95]. The AODV beacon interval is one second while OLSR uses two seconds.

10.4.1 Handover

In a first set of experiments we want to explore the handover quality. An exact measurement of the handover-time is difficult since it requires detailed information about the time the link disappears and the new routing entry is inserted. Moreover, the time between a device disappearing and appearing again does not tell anything about the fact whether a communication path between two nodes is restored or not. Restoring a communication path in an ad hoc network with multiple radios and a *vi*

10.4. SYSTEM EVALUATION

requires a lot more. It requires the *vi* to switch to the new physical interface and it may also require the route to be re-established. To measure all those effects we used the *ping* command to measure the lost packets. As *ping* triggers a packet transmission every second and as it is UDP based, there is almost a one-to-one relationship between the lost packets and the handover time. In order to take the impact of the routing module on the neighborhood database into account, we measured along three tracks. Once using AODV [96] representing the reactive class, once OLSR [94] as a proactive routing protocol and once without routing module at all. Measurements represent average values taken from 10 samples and the standard deviation is shown as well. Throughout the measurements we use *raw* to describe scenarios not including the virtual interface. On the contrary we use *vi/x* to indicate scenarios including the virtual interface with a *maxdiff* threshold of x. The results are shown in Table 10.3. Apart from measurements marked as 'priority driven', the priority table reflects the bandwidth proportions of the associated interfaces, therefore 802.11 always has the highest priority.

Let us look at the horizontal handover first. We trigger a route change by physically detaching the interface of a node. Under our setup we measured a packet loss of roughly 2 packets when the route changed from one hop to another, regardless of which routing protocol was used and whether a *vi* was included or not. But note that OLSR is not as stable as AODV (its standard deviation is higher). This is due to the fact that proactive routing protocols (like OLSR) do not react to route failures directly but broadcast their routing information on a fixed schedule instead. This behavior is even amplified in a pure Bluetooth network [6]. Here we observe a difference in packet loss between the two protocols of 10 packets (34.6 for AODV and 44.3 for OLSR). This makes sense since OLSR is running with a beacon interval of 2 seconds, while AODV uses only 1 second. It also explains the higher standard deviation of OLSR compared with the one of AODV. For the diagonal handover - we assume the Bluetooth connection already being established on the MAC level - a similar picture appears, OLSR is always a little behind AODV while being less stable. However, changing the route simultaneously with the MAC technology does not seem to introduce additional overhead in the AODV case. For the vertical handover (no route change, but interface switching only) we now also take the *maxdiff* threshold into account since it influences the agility in terms of the handover decision. Furthermore, we introduce *priority driven* handover. While in all other experiments a handover is triggered by the environment (in our scenario by physically detaching an interface), *priority driven* means the interface ranking is changed (using *victl setportpriority*) during a *ping* session. From the results given for AODV and OLSR in Table 10.3 we see that packet loss increases with increasing *maxdiff* threshold and almost disappears for priority driven handover. The former is reasonable because the bigger the *maxdiff* value, the more the priority policy gets enforced, and a pure priority driven MAC switching would not lead to any switching at all. The latter approves our idea of implementing MAC-switching transparent at the lowest layer possible. We also tested the impact of *maxdiff* values on priority driven handover. As expected, the smaller *maxdiff* gets the less stable the handover becomes. However, in our scenario a *maxdiff* value of 5-10 was sufficient to guarantee stable handover

---

[6]Note that after the route gets broken, a bluetooth inquiry is triggered in order to establish a new connection, which is a quite time consuming process.

84 CHAPTER 10. HETEROGENEOUS MOBILE AD HOC NETWORKS

Routing	From	To	Mode	Protocol	Throughput [Mbit/s]	Stddev
AODV	WL	WL	raw	TCP	5.029	0.0190
				UDP	6.019	0.008
			vi	TCP	5.019	0.0056
				UDP	5.984	0.066
	BT	BT	raw	TCP	0.4854	0.053
				UDP	0.4791	0.0677
			vi	TCP	0.4403	0.0538
				UDP	0.4830	0.05
	WL/BT	WL/BT	vi/1	TCP	5.018	0.0044
			vi/25	TCP	5.02	0.0044
			vi/200	TCP	5.021	0.0042
			vi/1	UDP	2.928	0.621
			vi/25	UDP	5.249	0.46
			vi/200	UDP	6.024	0.005
OLSR	WL/BT	WL/BT	vi	TCP	5.01	0.0105
			vi/1	UDP	2.034	0.667
			vi/25	UDP	5.154	0.538
			vi/200	UDP	5.816	0.224

Table 10.4: Throughput

while changing interface priorities. The very last part in Table 10.3 refers to vertical handover measurements without a MANET routing module. Here packet loss does not increase with increasing *maxdiff* threshold and no value is assigned to priority driven handover. This can be explained simply as follows. If no routing module exists, populating the Neighborhood Database (NBD) will occur as a side-effect of ARP (Address Resolution Protocol). After a link gets broken, the next upcoming ARP request, or to be concrete, the following ARP reply message creates a neighborhood entry within the NDB. Therefore, the handover time is a direct function of ARP request cycles, which is roughly 20 seconds in our example. The reason why priority driven handover doesn't work without routing module, is because no broadcast messages are received by the sending node. Typically, the NDB then contains only one entry.

10.4.2 Throughput

In the second experiment we measured the overhead produced by the virtual interface in terms of throughput. Again, our setup consists of two laptops, both equipped with Bluetooth (∼1Mbit/s) and a Cisco Aironet 350 Wireless LAN card (802.11b, 11Mbit/s). We measure both UDP and TCP throughput using the *iperf* [97] bandwidth measurement tool. Each experiment is repeated 10 times to compute the average and the standard deviation. From the table 10.4 we see that a pure 802.11 connection achieves a throughput of up to 5Mbit/s for TCP traffic and around 6Mbit/s for UDP. UDP is known to have a better performance in wireless networks since TCP's congestion control algorithm does not handle the unstable wireless link well. The same experiment, but including the *vi* on top, shows a comparable result. There

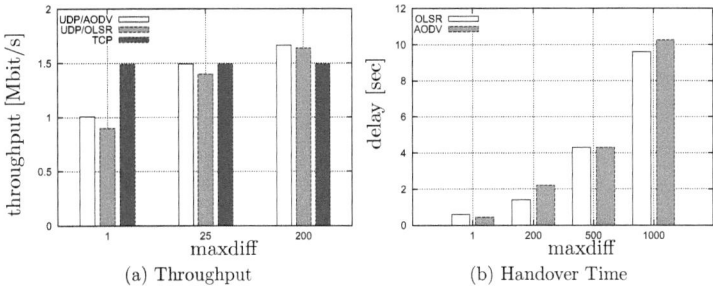

Figure 10.6: The *maxdiff* trade-off

seems to be almost no overhead produced by *vi*. This also applies to the Bluetooth case. An interesting result can be observed for the case where both machines are multihomed (WL/BT). While a low *maxdiff* threshold does not affect TCP traffic - neither for OLSR nor for AODV - it does so for UDP. The explanation is simple, TCP is acknowledgement based and therefore periodically refreshes the high priority entry within the NBD (WL in this case) during sending. UDP on the other hand does not know about acknowledgements and thus the entry within the NDB is only refreshed when it receives routing broadcast packets or ARP replies. This is causing the *vi* to use the lower priority interface occasionally which results in decreased throughput, especially if the *maxdiff* threshold is low.

10.4.3 Wrap-up

The measurements show that the system performs well. There is almost no overhead when using the additional *vi* on top of a physical interface. Generally, one can say that the *vi* performs better in combination with AODV than with OLSR. Except for the case of priority driven handover, the *vi* also works without a MANET routing module. However, a routing module increases the performance in terms of packet loss during handover. In the case of multiple physical interfaces there is a trade-off between agility in terms of vertical handover and throughput for UDP traffic (see Figure 10.6). A big *maxdiff* threshold is favorable for throughput but at the expense of handover agility, see Figure 10.6. This is due to the interface decision mechanism shown in Algorithm 10.1 that selects the outgoing interface for a given packet according to the interface priority and the up-to-dateness of neighborhood database entries. The weighting of these two factors is determined by the *maxdiff* value.

10.5 Discussion

The approach of using a virtual interface to integrate different wireless NICs with possibly different technologies has various advantages but also creates some problems.

One advantage – and the main criteria to choose such a solution – is that a virtual interface integrates well with the rest of the networking stack and requires almost no change on the application level. A disadvantage is that not all wireless NICs can be integrated into a *vi* that easily. As mentioned, some NICs do not allow for changing their MAC address. In principle, there is no problem with this since the neighboring database (NDB) will be able to determine the right outgoing interface for any given destination MAC address if it has received an incoming packet from the target device before. However, if the MAC address is exclusively associated with a remote NIC, there will only be one entry in the NDB for that given MAC address. In such a situation, Algorithm 10.1 will only have one option where to send a packet to. As a consequence, priority driven vertical handover will not be possible. Moreover, if there is no virtual MAC address per node and two nodes have multiple links to choose from (e.g. if both nodes have an 802.11 and a Bluetooth NIC) the ARP cache timeout determines how fast a node will recover from a link outage. One solution would be explicitely purge the ARP cache more often, but this will harm the performance since more ARP requests will be triggered.

10.6 Related Work

There is very little work on network abstractions for different media access technologies in ad hoc networks. In [98, 99], the authors describe the design and implementation of a virtual interface which allows to plug in multiple physical interfaces with possibly different channel configurations. However, their work does not consider multiple access technologies. Rather they focus on how to efficiently switch to the optimal channel at a given point in time. *VirtualWifi* is another approach based on virtualization [100, 101]. *VirtualWifi* abstracts a single wireless interface into multiple virtual interfaces. The basic idea is to switch the physical interface across multiple channels used by each virtual interface. There are other testbed works that abstract multiple real interfaces into a single virtual interface [102, 103], but again those approaches do not consider different access technologies.

In [104] the authors present, MR-LQSR, a MANET routing protocol for multradio wireless mult-hop networks. MR-LQSR is implemented as a virtual network adapter, similar to the virtual interface presented in this thesis. The protocol assigns weights to individual links based on the Expected Transmission Time (ETT) of a packet over a link and uses those weights when computing the route for a given packet. The difference between MR-LQSR and the *vi* is that MR-LQSR combines routing and multi-radio integration whereas we try to keep those two layers separated. The discussion about cross-layer software is a different one, however, incorporating more sophisticated link metrics would certainly be one way to improve the *vi*. More work on link metrics to increase the performance of wireless multi-hop networks can be found in [105]. To some extent, advanced link metrics are orthogonal to Algorithm 10.1 since the priorities of each phyisical interface in the *vi* and the *maxdiff* value can be changed at run-time using *victl* (see Table 10.1).

There is work on providing seamless access between multiple networks and access technologies in infrastructure-based WLAN networks. The swiss telecommunication company Swisscom developed a device which allows to seamlessly switch between WLAN, UMTS and GSM [106]. There are several initiatives to optimize mobility accross heterogeneous networks. The MOBOPTS working group within the IRTF (Internet Research Task Force) and the DNA (Detecting Network Attachment) working group within the IETF have been investigating ways to support optimized handover by using appropriate triggers and events from the lower layers. The IEEE 802.21 [107] working group is currently working towards a Media Independent Handover framework in a heteregeneous access environment. Suggestions to improve the handover time by taking advantage of IEEE 802.21 are given in [108]. A method using the Stream Control Transmission Protocol to facilitate seamless handover between wide-area cellular networks such as UMTS and WLAN is presented in [109].

There is also some work on integrating different access technologies in Personal Area Networks (PANs). ICON [110] offers seamless interoperability accross heterogeneous access technologies at mobile nodes in PANs. A Distributed Virtual Network Interface (DVNI) is proposed in [111] for device discovery, seamless handover and routing in PANs.

10.7 Summary

In this chapter, we have proposed an end-to-end communication abstraction that can be used in heterogeneous mobile ad hoc networks. Such networks are characterized by different MAC technologies used among the nodes. The proposed solution is based on a *virtual interface* (vi) approach. In this chapter we have shown how the virtual interface can be used to transparently integrate Bluetooth and 802.11 into one single IP-based MANET. Note that our solution is not bound to 802.11 or Bluetooth, but works together with any 802.x-compatible MAC Layer. The *vi* in combination with a MANET routing protocol supports multihoming, dynamic reconfiguration and node mobility. The experiments presented demonstrate the feasibility of the abstraction and its potential in building heterogeneous ad-hoc wireless networks.

Chapter 11
MAND: Mobile Ad Hoc Network Directory

In chapter 9 we expressed our belief that — in order for MANETs to become ubiquitous — MANETs will have to support existing Internet-based applications seamlessly. The problem we target in this chapter is how to implement a simple, efficient platform that supports standard network services over MANETs. The key observation behind the solution is that many networking services are based on mapping functionality. For instance, a DNS service maps domain names to IP addresses, or a SIP services maps a SIP URL to a TCP/IP endpoint where the user can currently be accessed. Hence, the focus of this chapter is how to provide an efficient key/value store that can be used to implement mapping functions which can then be used to implement basic network services in MANETs.

11.1 Problem Statement

11.1.1 The design space

A key/value store is a system that provides two basic functions. An insert operation to insert a key/value pair into the system, and a lookup operation to retrieve a value for a given key. For simplicity, we will use the term *tuple* as a short name for key/value pair in the rest of this thesis. In principle, there are three different options to implement a distributed key/value store. Either tuples are stored only locally (at the node that generates the tuple), or they are stored on individual nodes in the network, or they are stored everywhere. We refer to those three approaches as *local*, *hashed* and *replicated*. Any directory service in practice will implement one of those approaches or or a superposition of them. In the following we want to briefly study the various characteristics of each of those strategies to later motivate the approach we have taken in this thesis.

Performance: Let us first have a look on the performance of the *insert* and the *lookup* operation. In the *local* strategy insertion will be $O(1)$ and lookup will be $O(n)$, with n being the number of nodes in the network. This is because to look up a tuple, in the worst case it might be necessary to contact every other node in the network. On the other hand, inserting or updating a tuple becomes a local operation. This

90 CHAPTER 11. MAND: MOBILE AD HOC NETWORK DIRECTORY

is different with the *replicated* approach. There, insertion will be $O(n)$ and lookup $O(1)$. In the *hashed* strategy, the performance of the insert and lookup operations very much depend on which node is selected to store the tuple with a given key, and how fast this node can be discovered and contacted. DHTs – one class of key/value stores – achieve a performance of $O(\log n)$ for both insertion and lookup by storing tuples on nodes that are logically close to the key.

Consistency: Obviously, both the *local* and the *hashed* strategy allow nodes to update the tuples fast and atomically since there exists only one single instance of each tuple. Since network failures are common in MANETs, the *hashed* approach will have to be rated slightly worse because updating a tuple requires a connection to a remote node. A *replicated* approach, however, may result in a poor consistency leaving tuples in an inconsistent state. That is either because the distributed update of all replicas of a tuple cannot be handled atomically, or simply because some nodes cannot be reached and therefore their tuples cannot be updated. The latter may happen quite frequently in MANETs due routing failures or partial disconnections of certain network regions.

Availability: In terms of availability, a *replicated* strategy performs best. This is because even if some nodes turn offline or cannot be reached temporarily, looking up a tuple is still possible due to the replicas available on other nodes. Because of the replicas, a lookup for a key may even succeed after the originator of the tuple has already left the network. In some cases this is not the desired behavior, for instance if the tuple stored in the network contains information about a service provided by the originator of the tuple. If on the other hand, a *local* strategy is used, then the availability is low because other nodes will not be able to look up a tuple if they cannot contact the originator. This does not pose a problem if the tuple contains information about a local service. In this case it is sufficient for a tuple lookup to succeed whenever the service whose contact information is encoded in the tuple can be accessed. The availability in the *hashed* strategy is comparable to the one in the *local* case. However, again consider the case where tuples hold information about a service running on the node inserting the tuple into the system. Due to network partitions or routing failures, it may happen that some nodes cannot lookup the tuple including the service contact information, even if they could potentially access the service itself.

Memory: Using a *local* or a *hashed* strategy requires $O(t)$ memory space, while the memory requirement of a *replicated* approach is $O(t \cdot n)$, with t being the number of tuples and n the number of nodes in the network. This makes clear that, a full replication of tuples can only be an option in situations where both the number of tuples to be stored and the network size is small.

The various performance aspects of the three different architectures (local, hashed and replicated) are summarized in Table 11.1.

11.1.2 Discussion

Neither of the proposed solutions, *local, replicated* and *hashed*, performs well in all metrics. It is, for instance, impossible to guarantee data consistency and at the same time provide a fast lookup and high availability, as also stated in the CAP

	Update	Lookup	Consistency	Availability	Memory
Local	$O(1)$	$O(n)$	very good	bad	$O(t)$
Replicated	$O(n)$	$O(1)$	bad	good	$O(n \cdot t)$
Hashed	$O(\log n)$	$O(\log n)$	good	very bad	$O(t)$

Table 11.1: Performance vs Consistency vs Availability

theorem [112].

In *MAND*, we provide the user with the flexibility to trade different performance metrics on a per tuple granularity. Rather than choosing one fixed point in the design space, we turn the ability to operate in different modes into a feature of the system and a tunable parameter. This is particularly useful when supporting distributed network services in MANETs. For certain services, a fast response time is more important than a fast update time. This is the case, e.g., for a DNS service. DNS entries typically remain unchanged, but DNS requests are issued frequently in modern applications. For other services, it might be important to keep the stored tuples consistent. For instance location based services in ad hoc networks with high mobility. Yet another set of services require that the stored tuples are always fresh. One example is a presence service that allows to query the current online status of users. *MAND* provides the flexibility for all those services to operate at their preferred point in the design space. Moreover, *MAND* allows performance trade-offs to be made independently for each tuple that is inserted into the system.

11.2 System Model

The system model of *MAND* is organized along four design principles:

Fine-grain modes of operation to allow users to choose an optimal trade-off between different metrics such as, e.g., update performance, consistency, availability or lookup performance.

Routing integration to minimize the amount of traffic injected into the network, and to minimize the overall time needed to contact a higher layer service built on *MAND*.

Layering to guarantee independence of the underlying routing protocol.

Simplicity as the means to provide a robust and efficient software building block for implementing higher layer services in MANETs.

11.2.1 Architecture Overview

MAND runs as a user space process on nodes in the ad hoc network (see Figure 11.1)[1]. It provides two application interfaces, *insert*(tuple) and *lookup*(request), that operate on objects of two distinguished datatypes, *tuple* and *request*. If not explicitly mentioned we are using the terms *tuple* and *request* interchangeably for objects of the corresponding type.

---

[1] It is not a requirement that all nodes in the network are running an instance of *MAND*.

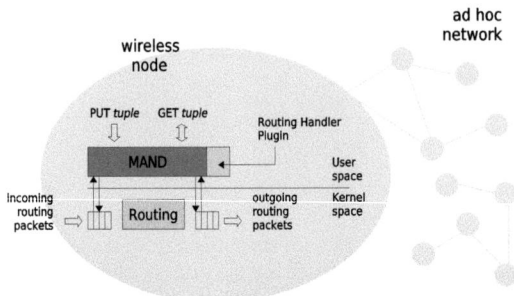

Figure 11.1: The *MAND* software architecture

The *insert* interface allows to pass a tuple to the system for storage and distribution. Each *tuple* associates a *key* to a *value*. Objects of type *tuple* are contained of six fields: the actual *key* and *value* forming the association, the *owner* identifying the node where the tuple has been inserted first, a *scope* specifying a network region (measured in hop distance from the owner) where to store the tuple, a *version* uniquely identifying the tuple together with the *key*, and a *lifetime* specifying for how long a *tuple* is going to be accessible on one particular *MAND* instance.

The *lookup* interface allows to lookup a tuple matching the constraints given by the *request*. A *request* has the same fields as a *tuple*, except that there is no *value* field. Any *tuple* returned by a *lookup* call must match the *key* specified by the corresponding *request*. The IP address of the node who issued the *request* is saved in the *owner* field. In contrast to a *tuple*, a *request* is identified by the triple (*key, owner, version*). And the *lifetime* specifies the time window during which matching tuples will be returned to the application layer.

The distribution of *tuples* and *requests* is done by piggybacking onto routing layer messages. The *MAND* user space process uses netfilter [113] to intercept routing layer messages right before and after they are processed by the routing module. By doing so, *MAND* makes sure that the routing protocol itself does not have to be modified or re-compiled. Research on MANET routing has a long history and there are plenty of different approaches to routing in MANETs. Depending on a particular situation, one protocol may outperform another. We do not constrain the choice for the optimal routing protocol by any higher level system. For this purpose, *MAND* uses the concept a *routing handler*. A *routing handler* is a plug-in that can be loaded by *MAND* at startup time. The plug-in encapsulates all the routing specific functionality such as piggybacking a tuple onto a given routing message, or isolating a tuple from a message. We will see in detail how the *MAND* system interacts with the *routing handler* in Section 11.3.

11.2.2 Inserting a tuple

When a tuple enters *MAND* on a node (through the *insert* interface), it gets processed in four steps (Algorithm 11.1). First we make sure the tuple is not an old tuple we have seen before (step 1, lines 1-7). A tuple is considered to be old

11.2. SYSTEM MODEL

Algorithm 11.1 insert(tuple)

1: old_tuple := NIL;
2: **if** tuple_store.contains(tuple.key) **then**
3:    old_tuple := tuple_store.find(tuple.key);
4:    **if** tuple.version $\leq$ old_tuple.version **then**
5:      return;
6:    **end if**
7: **end if**
8: tuple.scope := tuple.scope - 1;
9: **if** old_tuple $\neq$ NIL **then**
10:    tuple_store.update(old_tuple, tuple);
11: **else**
12:    tuple_store.add(tuple);
13: **end if**
14: **for all** request $\in$ request_store **do**
15:    **if** request.key = tuple.key **then**
16:      reply(request, tuple);
17:    **end if**
18: **end for**
19: **if** tuple.scope $\geq$ 0 **then**
20:    routing_handler.process_tuple(tuple.clone());
21: **end if**

if its version is lower or equal the version of an already stored tuple. In this case we immediately abort the processing. Otherwise, we decrement the scope of the tuple by one and add it to our local store. If there exists already a tuple with the same key, but a with a lower version, we may replace the existing tuple (step 2, lines 8-13). *MAND* stores *tuples* and *requests* in two separate data structures named *tuple_store* and *request_store*. The *request_store* is queried in the third step to check if there are any pending application requests for the given tuple. In contrast to the *tuple_store*, the *request_store* may contain several requests with the same *key*. For each pending *request* that is found, a corresponding reply message is sent to the application, including the newly received tuple (step 3, lines 14-18). The last thing to be done is to forward the tuple if necessary. If the decremented scope of the tuple is still greater than zero then a copy of the tuple is delivered to the *routing handler* (step 4, lines 19-21). The *routing handler* will append the tuple to a routing message that is intercepted just before the message was going to leave the network interface. Depending on whether the message is a broadcast or a unicast message it is received by many or just one neighboring node. There, the routing message is intercepted by *MAND* right before it is processed by the routing module. *MAND* then passes the message to the *routing handler*, where the attached *tuple* will be isolated from the message. At that point on the remote node, the tuple will be processed according to Algorithm 11.1, just as it would have been received through the application interface. In that sense, a tuple may travel across the network until its scope becomes smaller than zero.

As mentioned before, tuples are automatically removed once their *lifetime* expires. However, in order to avoid tuples being distributed in loops, *MAND* makes sure that

Algorithm 11.2 lookup(request) : tuple

1: request.scope := request.scope - 1;
2: is_fresh := true;
3: **if** request_store.contains(request.key) **then**
4:     old_request := request_store.find(request.key);
5:     **if** old_request.version ≥ request.version **then**
6:         is_fresh := false;
7:     **else if** old_request.owner = request.owner **then**
8:         request_store.add(request);
9:     **end if**
10: **end if**
11: **if** tuple_store.contains(request.key) **then**
12:     tuple := tuple_store.find(request.key);
13:     **if** is_local(request.key) **then**
14:         reply(request, tuple);
15:     **else if** is_fresh **then**
16:         routing_handler.process_reply(request, tuple);
17:     **end if**
18: **else if** is_fresh ∧ request.scope ≥ 0 **then**
19:     routing_handler.process_request(request);
20: **end if**

every tuple is stored for at least l_{min} seconds. Tuples with a *lifetime* of less than l_{min} will be cached for additional $l_{min} - lifetime$ seconds. During that time, the tuple is considered to be in the *invisible* state, meaning that it cannot be looked up by any node in the network. Invisible tuples prevent the distribution of tuples entering the system with the same key.

11.2.3 Looking up a tuple

When a *request* enters a node through the *lookup(request)* interface, it gets processed as follows (Algorithm 11.2). First, its scope value is decremented (line 1). Second, the system checks whether there is a already a matching request in the local *request_store*. Remember, that two requests match only if they have a common key, version and owner. If no matching request was found, the new request is added to the *request_store*. Should there exist an old request with the same key and owner, but with an older version, then the old request will be replaced (lines 3-12). In both cases we say the received request is *fresh*. In a third step, we check whether we can find a tuple for the requested key in the local *tuple_store* (lines 13-14). If a matching tuple can be found, and the request was issued from a local application, then a reply message will be sent to the application including the found tuple (lines 15-16). If, however, the request has been received from the network, and the request is *fresh*, then the found tuple together with the request is delivered to the *routing handler* (line 17-18). The *routing handler* makes sure the tuple will be sent back to the owner of the request in an efficient way. More details on the internals of the *routing handler* are presented in section 11.3. If, after all, no tuple matching the input request can be found in the local *tuple_store*, then the request is passed to the *routing handler* to be forwarded further (lines 17-19). Note that, as with the tuple, a request is only forwarded if it is *fresh* and its scope is greater or equal than zero. Generally, the forwarding of a request results in the same set of actions as the forwarding of a tuple. The *routing handler* attaches the request onto an existing routing message. On a neighboring node, the request is detached from the message by the *routing handler* and it will again be processed as described in Algorithm 11.2.

11.3 Routing handler

So far we have seen how the *routing handler* is involved while storing or looking up a tuple. In general, the *routing handler* plug-in makes it possible for $MAND$ to take advantage of the routing protocol's mechanisms to forward and lookup tuples, while at the same time not being dependent on the actual routing protocol used. The difficulty is to design a minimal interface between the *routing handler* and $MAND$ which is powerful enough to support most types of routing protocols.

We have modelled the *routing handler* as a passive component that is invoked by $MAND$ using the following five function calls:

- *parse(routing_message)* is invoked whenever a new routing message is intercepted. The *routing handler* must parse the routing specific part of the message and provide access to all extra data attached to a routing message.

- *process_tuple*(tuple) is called by $MAND$ whenever a tuple needs to be forwarded according to Algorithm 11.1. The *routing handler* may store the *tuple* temporarily until a routing message becomes available for piggybacking.

- *process_request*(request) is called by $MAND$ whenever a request needs to be forwarded according to Algorithm 11.2. Similar to the *process_tuple* function, the *routing handler* is supposed to piggyback the *request* as soon as a routing message becomes available.

- *process_reply*(request, tuple, routing_message) is called whenever a request received from a remote node matches a tuple in the local store. $MAND$ expects the *routing handler* to make sure the matching tuple is sent back to the owner of the request (see Algorithm 11.2).

- *piggyback*(routing_message) is called by $MAND$ before a routing message is going to leave the NIC node. The *routing handler* may use the *routing_message* to piggyback *tuples* and *requests* it has received during previous *process_tuple* or *process_request* function calls.

A *routing_handler* for a concrete routing protocol will have to provide implementations for those five functions. Thereby, a *routing handler* should try to exploit the routing mechanism of the corresponding routing protocol. There is a vast amount of different approaches to perform routing in ad hoc networks. However, most existing routing protocols either belong to the class of *pro-active* or to the class of *reactive* routing protocols. *Pro-active* protocols maintain fresh lists of destinations and their routes by periodically distributing routing tables throughout the network. *Reactive* protocols find routes on demand by flooding the network with route request packets

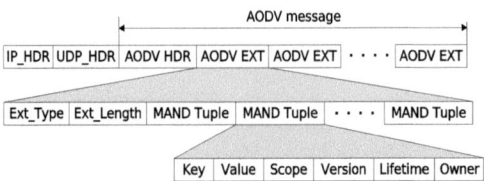

Figure 11.2: Piggybacking MAND tuples onto an AODV routing message

whenever there is an application request. We have implemented two examples of a *routing handler* for both a *pro-active*, and a *re-active* routing protocol.

11.3.1 OLSR

OLSR is a *pro-active* routing protocol. The OLSR *routing handler* makes use of the fact that OLSR periodically advertises its routing table. More precisely, the *routing handler* temporarily stores *tuples* it has received through the *process_tuple* interface, and piggybacks them onto OLSR route advertisement messages whenever the *piggyback* function is invoked. The actual piggybacking is done by storing the tuple information inside an OSLR extension message as specified in [114]. OLSR is purely pro-active and does not provide explicit route request functionality. Hence, the OLSR *routing handler* implements the *process_request* interface by piggybacking *requests* onto OLSR route advertisements, just as they would be regular *tuples*. Remember that, whenever a remote request matches a tuple in a remote *tuple_store*, the *process_reply* function of the *routing handler* is called. The OLSR *routing handler* implements the *process_reply* interface by sending a unicast UDP packet including the matching tuple to the owner of the request. The format of the message sent is the same as the one of messages exchanged between a local application and *MAND*.

11.3.2 AODV

AODV is a *re-active* routing protocol. Thus, there are no periodic route advertisement messages that could be used to piggyback *tuples*. Instead, the AODV *routing handler* piggybacks *tuples* onto route request and route reply messages that are issued by AODV on behalf of application requests. Additionally, the routing handler may also piggyback *tuples* onto AODV *HELLO* messages if available. The RFC [93] defines those messages as optional, but in practice most implementations do use *HELLO* messages to increase the robustness of the protocol. Piggybacking *tuples* is done by storing the *tuple* information inside an AODV extension (see Figure 11.2), as specified in [96]. The maximum size of an AODV extension is limited by 255 bytes due to an 8 bit length field in the extension header. Therefore, the AODV *routing handler* fragments tuples with a larger size into multiple data blocks of smaller size and spreads them across multiple AODV extensions.

One strength of AODV is its ability to discover routes on demand. AODV adopts an expanding ring search technique where the network search region for a route request is doubled in each phase until either the target node is found or the entire network is reached. The AODV *routing handler* exploits this mechanism when being asked to forward a *request* for a *tuple*. Upon receiving a *process_request* call, the handler temporarily stores the submitted *request* and triggers an AODV route request. The newly triggered route request message is intercepted by *MAND* immediately and passed to the handler's *piggyback* function where the previously stored *request* is attached. Triggering a route request can be done without interacting with the routing protocol directly, for instance by transmitting a UDP packet to an IP address that does not exist in the ad hoc network. *MAND* manages a pool of configurable IP addresses to be used for this purpose.

Any *routing handler* must implement the *process_reply* interface to guarantee that tuples matching remote requests are actually sent back to the originator of the request. The AODV *routing handler*, when being invoked by a *process_reply* function call, stores the matching tuple and triggers a route reply. The triggered AODV route reply message is intercepted immediately by *MAND* and passed to the handler's *piggyback* function where the previously stored tuple is attached. Triggering a route reply, given the route request, is done simply by overwriting the route request's destination to be the local IP address. The altered route request message appears to AODV like a route request for the local node and makes AODV respond with a route reply message.

Overall, the AODV *routing handler* entirely maps a remote tuple lookup to the AODV route-request/route-reply procedure. By doing so, it is guaranteed that for each successful remote *MAND* lookup request a route will be established to the node hosting the requested *tuple*. This is of particular importance if the requested tuple contains information about a service running on the host node. Obviously, the time spent to remotely look up a tuple is saved later when accessing the service since no route establishment is needed anymore.

It is important to note that neither the AODV nor the OLSR *routing handler* interact with the routing protocol directly. The advantage is that the routing protocol does not have to be modified or re-compiled. This is a necessary requirement when using *MAND* in practice since users may not have control over the routing protocol implementation.

11.4 Interface Semantics

In section 11.1 we discussed the various trade-offs involved in implementing a distributed directory service. In *MAND* the behavior is controlled using the two fields *scope* and *lifetime*. These fields are available in both *request* and *tuple* and thus are exposed to the application level API (*insert* and *lookup*).

11.4.1 Controlled insertion of tuples

The advantages and disadvantages of a *local* versus a *replicated* insertion strategy have been demonstrated in section 11.1. Using the *scope* parameter of a tuple, a user of *MAND* can configure any point of operation between *local* and fully *replicated*.

An application may use the *lifetime* parameter of a tuple together with the tuple update frequency adapt to the dynamics of the network. A tuple can be kept persistent in the network if it is periodically updated with a newer version before the

98 CHAPTER 11. MAND: MOBILE AD HOC NETWORK DIRECTORY

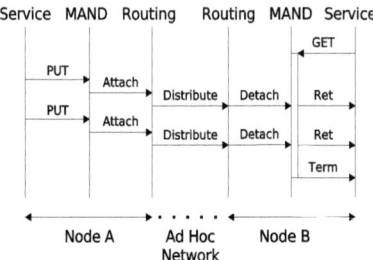

Figure 11.3: Example 3: Event/Subscription. Tuple request with **scope** > 0 and **lifetime** > 0. Tuple insert with **scope** $= 0$ and **lifetime**> 0

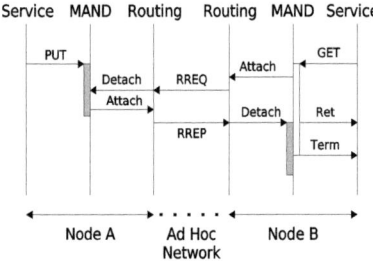

Figure 11.4: Example 2: Remote lookup. Tuple request with **scope** > 0 and **lifetime** > 0. Tuple insert with **scope** $= 0$ and **lifetime**> 0

lifetime expires. A short *lifetime* could be useful to avoid stale and outdated tuples in a MANET that is exposed to network partitions. Network partitions may prevent a node from receiving the latest update for a tuple. With the *lifetime* being short together with an appropriate update frequency, it is guaranteed those tuples will be either be up to date or removed quickly. On the other hand, a long tuple *lifetime* is used if the network topology does not change often or if tuple updates are rare. If the *lifetime* of a tuple is set to zero, a tuple will not be stored in the *tuple_store* of the node. It will, however, be forwarded if its scope is large enough. We call those tuples *events*, since they trigger reply messages for pending local requests (see Algorithm 11.2). Figure 11.3 illustrates the distribution of an event. Node B in Figure 11.3 has issued a request with a large *lifetime* to make sure it is notified whenever an event is received for the specified *key*. Since the event is not stored in *MAND* on intermediate nodes, any subsequent event for the same key will be considered as *fresh* and therefore will be forwarded (see Algorithm 11.1). This allows for events sharing a common key to be triggered by different nodes in the network.

11.4. INTERFACE SEMANTICS

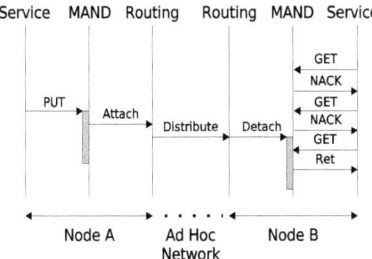

Figure 11.5: Example 1: Polling. Tuple request with **scope**=0 and **lifetime**=0. Tuple insert with **scope** > 0 and **lifetime**> 0

11.4.2 Controlled lookup of tuples

When searching for a tuple using the *MAND lookup* interface, an application may use the *scope* parameter to limit the network search region in terms of hops. In practice, a *lookup* with a *large* scope is often combined with a rather small scope during tuple insertion. The *scope* value used during an *insert* or *lookup* operation mirrors the frequency in which those operations take place. One typical use case for requests with a large *scope* are situations where searching is a rare but an important operation. In this case we shift some of the communication cost incurring during tuple insertion to the lookup operation. When using an on-demand routing protocol, we have seen in section 11.3.2 that the lookup costs might be compensated later because no additional route establishment will be necessary. Figure 11.4 illustrates a remote tuple lookup. There, the request is issued with a *scope* > 0 whereas the tuple is inserted with *scope* = 0. Note that the request has a lifetime (white bar in Figure 11.4) specifying the time window during which matching tuples will be accepted. There are also good reasons to issue tuple requests with a small *scope*. A *scope* of zero for instance can be used to make sure the search query is applied only to the local *MAND* instance.

An application may use the *lifetime* parameter of a tuple request to control the time *MAND* waits until it receives tuples with the requested key. In some cases, applications may not want to wait at all. Then the best they can do is to issue a request with zero *lifetime* and zero *scope*. If a matching tuple is found, a response including the matching tuple is sent back to the application. If no matching tuple is found, the request expires immediately causing a negative acknowledgment (NACK) to be sent back to the application. In either case, the tuple request call is non-blocking, which is a requirement in many applications. In Figure 11.5, the application periodically performs a local lookup with *scope* and *lifetime* set to zero. In the example, the first two requests are replied with a NACK until the third request successfully returns the requested tuple. The tuple has been distributed in the network due to an *insert* of a tuple with *scope* > 0 on a remote node. Note that, while the tuple was distributed in the network, every node who is forwarding the tuple will also store a copy of it in its local *tuple_store*. Figure 11.5 also illustrates the lifetime of a tuple (grey bar). The lifetime starts on every node at the moment the tuple is received. Hence, different

100   CHAPTER 11. MAND: MOBILE AD HOC NETWORK DIRECTORY

copies of the same tuple in the network do not expire exactly in sync.

11.5 Evaluation

We have implemented a prototype of *MAND* for Linux. The system completely operates on the user level, only requiring *netfilter/libipq* support from the kernel. Most kernels do have *libipq* support turned on by default. If not, the corresponding kernel modules can be added on-demand. We have tested our *MAND* prototype on both 2.4 and 2.6 kernels and on different hardware platforms ranging from laptops to handhelds. In this section we evaluate *MAND* using our prototype implementation. The experiments are divided into two parts. The first part studies the network coverage and availability when inserting a tuple in different scenarios with different HELLO intervals. In the second part, we look at the propagation time of tuples under different levels of background noise. The main objective of the experiments is to evaluate *MAND* versus the different performance metrics discussed in section 11.1. We will show what is the availability of a tuple once it has been inserted in *MAND*, or how fast a tuple is propagated in the network.

11.5.1 Evaluation setup

Our experimental testbed consists of 27 wireless nodes. Each node is a Soekris 4826 box with 266M Geode SC1100 processor, 128M RAM, and a 2.4GHz IEEE 802.11 b/g wireless NIC. The nodes are deployed in an indoor environment. They all connect to a central control server with a wired ethernet that we use for logging experimental data. The server also runs an ntp[2] service used by the nodes to synchronize their clock. The transmission range of the wireless NICs is effectively reduced by connecting the wireless NIC to a low-gain antenna through RF attenuators. All nodes are running Debian Linux version 'Etch' using a 2.6.18-6 kernel. Our experiments are carried out for both AODV and OLSR. We are using unmodified Linux implementations of both protocols[3,4].

11.5.2 Network coverage

The first experiment studies the network coverage of a tuple after an *insert* operation. At one corner of the network we insert, at intervals of 10 seconds, 30 tuples with the same key but with an increasing *version* number. Each tuple is inserted with a sufficiently large *scope* which schedules the tuple to be replicated on the entire network. In each round we also log the time just before the tuple is inserted. All other nodes in the network wait for tuples to be received locally by issuing a *lookup* request for the common key with a large *lifetime* and a *scope* of zero. Whenever a tuple is received by a node the time is logged.

Figure 11.6 shows, for each of the 30 rounds, what part of the network is covered by the tuple at a given time (each line corresponds to one round). The x-axis of the Figures corresponds the time passed after the tuple was inserted. The y-axis refers

---
[2]Network Time Protocol, http://www.ntp.org/
[3]AODV, http://core.it.uu.se/core/index.php/AODV-UU
[4]OLSR, http://www.olsr.org/

11.5. EVALUATION

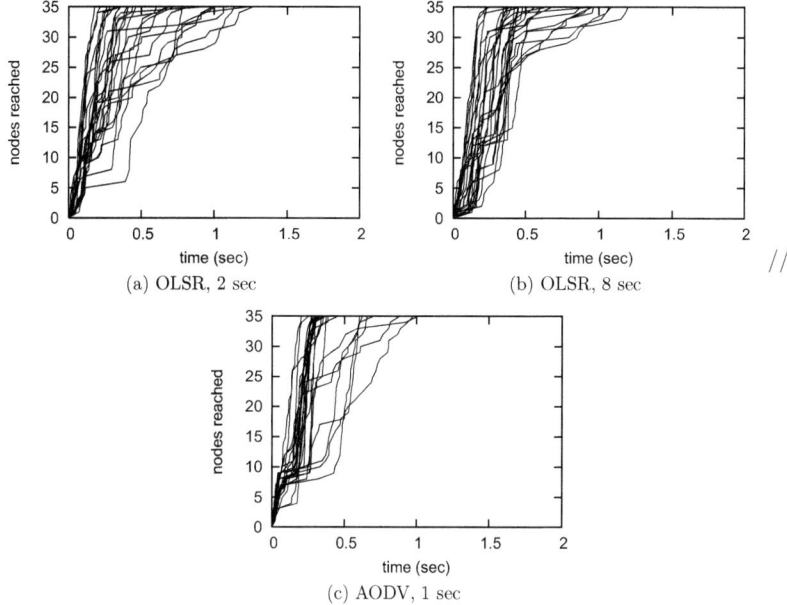

Figure 11.6: Network coverage for different Hello intervals

to the number of nodes having successfully received a tuple. The experiment has been repeated three times, once using OLSR with a *HELLO* interval of 2 seconds (Figure 11.6a), once using OLSR with a *HELLO* interval of 8 seconds (Figure 11.6b), and once using AODV with a *HELLO* interval of 1 second (Figure 11.6c). As one can see from the Figures, every tuple eventually reaches every node in the network. Interestingly, the time until a tuple has covered the entire network is comparable in all three scenarios. This might be surprising because of the different *HELLO* intervals that were used. However, it turns out that the coverage behavior, at least for OLSR, does not so much depend on the actual *HELLO* interval. This is because the *HELLO* messages only take a small part of all OLSR routing messages that are used by *MAND* for piggybacking. Figure 11.6c demonstrates that AODV is competitive in terms of network coverage if the *HELLO* feature is turned on.

11.5.3 Propagation time

In this experiment we study the propagation time of a tuple with respect to hop distance. The setup matches the one in section 11.5.2, but this time we also log the number of hops a tuple has passed when it is received. Figures 11.7a-c show the correlation between propagation time and hop distance for OLSR with a 2 second *HELLO* interval. We have conducted the experiments for three different levels of background noise. The background noise is generated by injecting additional tuples

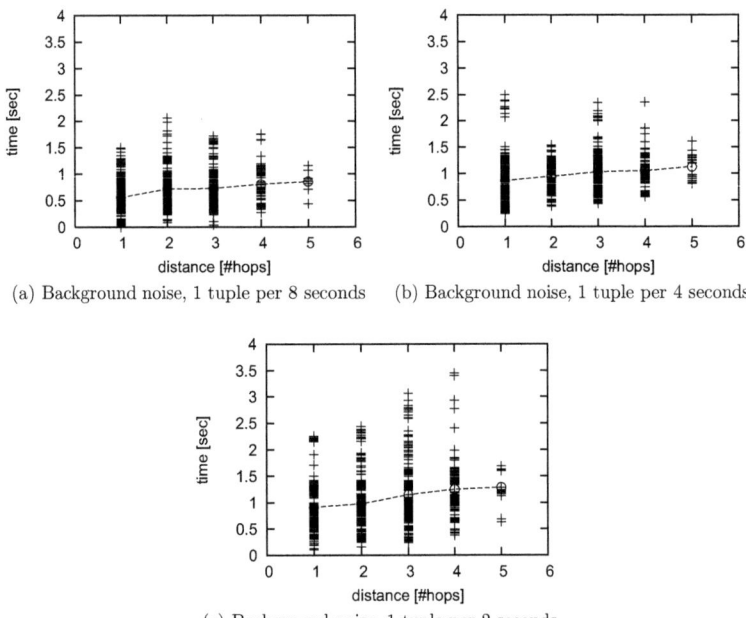

Figure 11.7: Tuple propagation in OLSR with under tuple generation rates

on nodes performing the requests. Each point in Figures 11.7a-c corresponds to the reception of a tuple 'x' hops away from the originator. The line shown in each Figure refers to the average propagation time. For a $HELLO$ message interval of p seconds the maximum propagation time of a tuple traveling k hops is $2 \times k$ seconds.

A first observation from Figures 11.7a-c is that the measured propagation time stays significantly below the worst case value. This is because the tuples propagate with a high redundancy. There is set of alternative paths between any node in the network and the node at the corner where the insertion of the tuple takes place. The propagation time of a tuple along a given path is determined by the length of the path and its schedule. Some paths may have a slow schedule. This is if a tuple, upon reception at a node, always just misses the next routing message. Some paths, however, might really have a fast schedule. This is if the reception of a tuple on a node is immediately followed by the transmission of a routing message. The tuple arriving first at a given receiver node is the one where the schedule and the path length did result in the fastest propagation time among all the different paths.

A second observation from Figures 11.7a-c is that the avarage propagation time increases with the background noise. This makes sense since tuples will have to stay in the queue for a longer time before they can be piggybacked. It can also be seen that, the more tuple background noise we have, the higher the variations in the tuple arrival rate.

11.5. EVALUATION

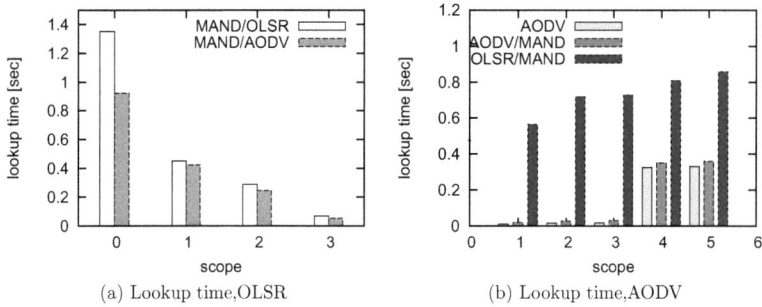

(a) Lookup time, OLSR   (b) Lookup time, AODV

Figure 11.8: *MAND* evaluation on the testbed: Lookup time

11.5.4 Lookup time vs. *scope*

In this experiment we study the time needed to remotely lookup a tuple. The setup matches the one in section 11.5.2, except that we insert the tuple with four different replication levels (*scope*=0,1,2,3). A *scope* of zero means that the *tuple* is not replicated at all. A *scope* of 3 means that the *tuple* is replicated on all nodes that are reachable over 3 hops. Figure 11.8a shows the lookup time in each of those cases. As we can see, the lookup time decreases with the replication level of the *tuple*. This is because the higher the replication level, the more likely it is that a tuple can be retrieved from a closeby node.

11.5.5 Lookup time vs. Hop Distance

In this experiment we relate the *tuple* lookup time to the hop distance the tuple is away from the requestor. For this purpose, a tuple with a **scope** of 0 and a **lifetime** of 20 is inserted into *MAND* on one node of the network. Another node k hops away makes a *MAND* lookup call for that given tuple using a **scope** of 10 and a **lifetime** of 20. The lookup time refers to the time from the moment the lookup call is made until the requested tuple is returned.

Figure 11.8b shows the measured lookup time when using AODV and OLSR as the ad hoc routing protocol. As a reference, we also show the AODV route establishment time.

A first observation from Figure 11.8b is that the lookup time when running AODV is faster than the lookup time when running OLSR. This makes sense since the *routing handler* for AODV integrates *tuple* requests directly with AODV's RREQ/RREP mechanism, while for OLSR *tuple* requests are piggybacked just as regular *tuples*. A second observation from Figure 11.8b is that looking up a tuple using *MAND* and AODV only takes a few milliseconds longer than a simple AODV route request for the same hop distance. Therefore, the price for piggybacking is minimal. The gap in time between three and four hops in Figure 11.8b is due to AODV's *expanded ring search* technique [93], which we have discussed in section 11.3.2.

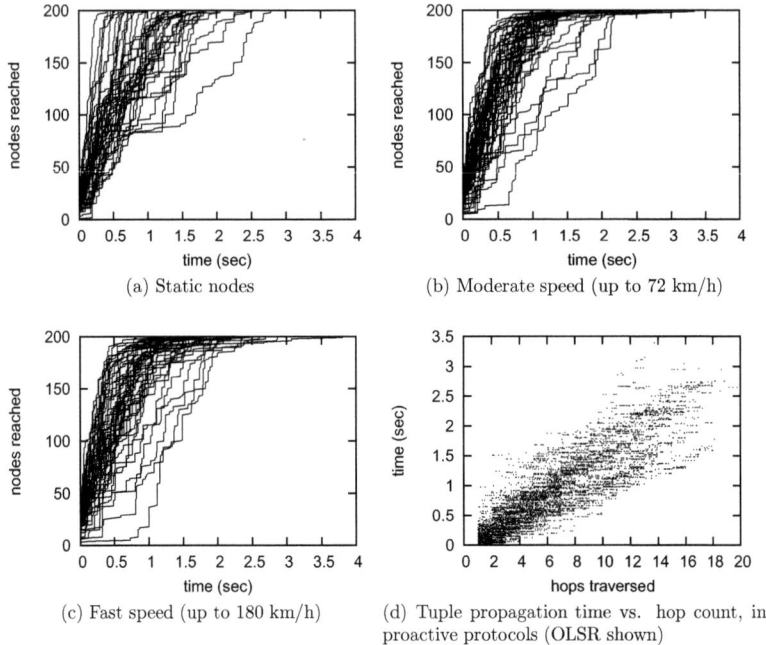

Figure 11.9: Tuple insertion over the OLSR protocol. Demonstrates the dynamics of tuple propagation in proactive protocols, by means of HELLO messages, for static and mobile scenarios.

11.6 Simulation

In addition to deploying *MAND* on a real network, we also implemented it and simulated it on NS2 [72]. Our basic configuration is a set of 50 nodes spread uniformly at random across a $1500m \times 300m$ plane. In order to test our system in a larger scale, we also consider a configuration of 200 nodes spread across a plane of dimensions $3000m \times 600m$, to keep the node density constant. We used the default AODV implementation included in the NS simulator, and the Universidad de Murcia UM-OLSR [115] implementation for NS.

11.6.1 Inserting Tuples

Propagation Speed

We carried out a number of experiments to explore the dynamics of tuple propagation in proactive protocols. The first observation is that, unless tuples are discarded due to queue overflow under congestion (see Section 11.6.1 below), they reach *all* nodes of the network. In our experiments we never observed a tuple propagation being hindered by

11.6. SIMULATION

the occasional packet collisions that occurred. This is due to the inherent redundancy in tuple propagation in proactive protocols, comprising numerous intermingled paths.

To observe the tuple propagation speed in proactive protocols, we ran *MAND* over a simulated network of 200 nodes running OLSR, introduced 50 tuples originating at randomly selected nodes, and kept track of the times at which they reached other nodes. To keep our measurements unbiased, we let tuples propagate at non-overlapping periods, and induced no traffic to the network. That is, only hello messages were being exchanged by the routing protocol.

Figure 11.9 plots the number of nodes that have received a tuple as a function of the time elapsed since the first transmission of the tuple, for a static network and for two instances of mobile networks. All three scenarios demonstrate that *all tuples* are spread *fast* and reach *all nodes*.

It should be emphasized that node mobility does not affect the operation of *MAND*. Quite on the contrary, it has a positive effect on it, accelerating the propagation of tuples. Figures 11.9(b) and (c) (nodes moving in random directions and at random speeds up to 72 km/h and 180 km/h, respectively) reveal lower propagation times compared to the static network of Figure 11.9(a). This is a particularly interesting result, and the behavior of our system under mobility is of particular importance to our work. Figure 11.9(c) shows the correlation of hops traversed and the respective time for a static network running OLSR. As expected, this is on average a linear relationship.

Finally, it should be noted that we carried out the same experiments also over AODV, having configured it to send HELLO messages. Setting equal periods of HELLO messages in the two protocols leads to indistinguishable results.

Tuple Forwarding History

In *MAND* tuples are forwarded whenever their scope is greater or equal than zero and no existing tuple with the *key* is already available in the local store. In section 11.2.2 we discussed that every tuple is stored for at least l_{min} seconds to avoid tuples being propagated in a "ping-pong" fashion if their actual *lifetime* is small. We call l_{min} the the *tuple forwarding history*, which is the time we cache tuples that were recently forwarded. A long history would serve our goal of redundancy avoidance, but would consume more resources, both in terms of node memory as well as node processing. A very short history could prove inadequate in certain networks.

In this section we perform a series of experiments to determine the optimal history length. Figure 11.10 shows the average number of times a tuple is forwarded by each node, as a function of the history length (in seconds). For very short history lengths, nodes "forget" tuples they have recently forwarded, and are willing to forward them again upon subsequent reception. In fact, recycling of tuples would go on for ever, unless tuple propagation was limited by the scope field.

In Figure 11.10 we consider a network of 50 nodes, and tuples are inserted with a scope of ten. We can see that for a history length of two seconds or longer, tuples are forwarded exactly once by each node, which is the desired behavior. Clearly, this crucial history length depends on the network topology and the cycles that can be formed. A good design should ensure that an adequate history length is allocated per node, to avoid unnecessary tuple forwarding overhead and congestion.

CHAPTER 11. MAND: MOBILE AD HOC NETWORK DIRECTORY

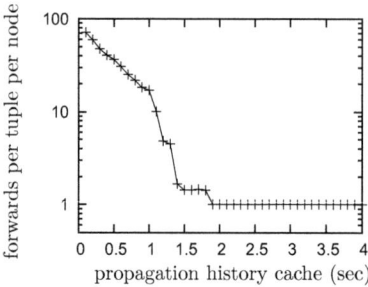

Figure 11.10: The effect of tuple caching time on redundant tuple retransmissions

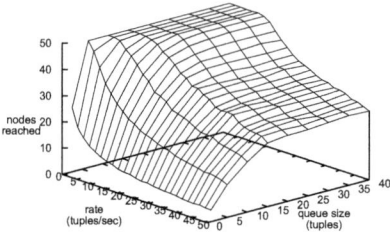

Figure 11.11: The effect of *tuple generation rate* and *queue sizes* on the number of nodes reached.

Tuple Insertion Rate and Queuing Issues

It is interesting to test what tuple insertion rate our system can sustain. As a finite number of tuples can be piggybacked on a single hello message, tuples will start queuing if they are generated at high rate. Queues also have a limited size, which leads to packets being dropped once queues are full, having a negative effect on the hit ratio. We carried out a number of experiments to test the limits of *MAND* with respect to the rate of tuple arrivals. Figure 11.11 explores the space of tuple arrival rates and queue sizes on a network of 50 nodes. We see that for very small queue sizes, even slow tuple arrival rates cause queue overflows, resulting in tuples reaching on average fewer than 50 nodes. For larger queue sizes, *MAND* can sustain higher tuple arrival rates, with all tuples reaching all 50 nodes.

11.6.2 Looking up Tuples

In this section we turn our focus on tuple lookup for *reactive* protocols. Remember thatn, when looking up a tuple, a node triggers a RREQ message, unless the tuple is already available in the local storage. If the RREQ message reaches a node that has the requested tuple, that node replies with a RREP, and piggybacks the requested

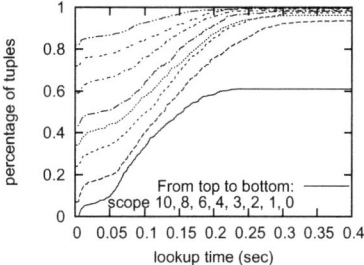

Figure 11.12: CDF of tuple lookup times for AODV. Each line corresponds to tuples spread with a different scope prior to their lookup.

tuple on it.

Clearly, the success rate and the lookup times depend on the extent at which the tuple in question is spread. In the extreme case that the tuple lays exclusively on its originating node, that specific node has to be reached by the requester's RREQ, and the single RREP sent back has to reach the requester. This may induce higher delays and more sensitivity to packet loss compared to the case of an abundantly replicated tuple, that would result in multiple RREP messages being send back to the requester. The other extreme, of course, is a tuple spread to the whole network, in which case the requester has a copy of the tuple in its local storage.

Figure 11.12 shows the CDF of the lookup times in a network of 200 nodes in a 3000m by 600m plain, for tuples initially spread with scopes from 0 to 10. Note that for non-replicated tuples (scope 0), lookup times are longer, and not all requests are successful. As the replication level increases, we observe lower lookup times, and higher success rates of lookups.

11.7 Related Work

In the past, there have been various different approaches to key based search in MANETs. Lindemann and Waldhorst showed how to apply epidemic dissemination to implement a distributed directory and introduced an analytical performance modeling approach [116, 117]. They presented a general-purpose distributed lookup service for mobile applications, denoted Passive Distributed Indexing (PDI). PDI stores index entries in form of (key, value) pairs in index caches maintained by each node. Index caches are filled by epidemic dissemination of popular index entries. Their approach, however, assumes that nodes are moving and exchanging cache entries whenever they are physically close.

Another approach is to provide efficient lookup facilities rather than replicating entries. Cell Hash Routing (CHR) [76] is a specialized ad-hoc DHT. CHR uses position information to construct a DHT of clusters instead of organizing individual nodes in the overlay. The limitations of this approach is that maintaining the cluster can be very expensive under node mobility. Moreover, the approach is not generic enough since it requires position information of the nodes. Cramer and Fuhrmann [118] propose the Chord based Proximity Neighbor Selection Strategy (PNS_CHORD). Here, the nodes are connected to their logical successors on the ring and through logical shortcuts to further nodes as usual in Chord [119]. However, because the construction of routing tables is absed on physical proximity, it may happen that the logical path pursued by a request traverses the same node multiple times. This problem has been tackled in [79] where the authors propose an adaptive learning process to prevent such loops. However, as all other DHT based approaches, also [79] is very inefficient when it comes the number actual messages that are transmitted per lookup request. This is because, transmitting a packet along an overlay may cause route requests on every hop. Ekta [75] and MADPastry [77] integrate key lookup facilities with the AODV routing protocol. And pucha et at. [120] implements Pastry [121] on top of Dynamic Source Routing (DSR) [122]. Rowstron et al. propose VRR [123], a routing protocol including logical DHT-like addressing. The problem all these approaches have is that they rely on a particular routing protocol which is not useful in practice.

The problem of key based search has similarities with service discovery and name resolution in MANETs. One approach to the latter is to let service information be maintained on a per-cluster basis [124, 125]. However, electing and maintaining specialized nodes is difficult and expensive due to the dynamic nature of MANETs. In fact, it has been shown that in terms of transmitted messages, a fully distributed service discovery outperforms the per-cluster approach, especially in mobile scenarios [125].

Another related problem to key based search can be found in geographical routing. Geographical routing algorithms assume the availability of geographical location information of nodes (e.g., via GPS or other techniques). Using this information nodes make packet forwarding decisions based on the packet destination location and neighboring nodes locations. To discover the location information, geographical routing makes use of so called *location services*. In principle, the *location service* maps a unique ID of a node, to the current location of the node. One approach to provide location services is to divide the network area into a flat grid of squares, and then store the location information of a node on all other nodes residing inside a randomly chosen square, dependent on the ID of the node and the coordinates of the square[126, 127, 128]. The disadvantage of this approach is that the routing costs to query the location of a node grow proportional to the network size, even if the two particapting nodes are geographically close to each other. To improve the performance of location services hierarchical directory structures were suggested [129, 130, 131, 132]. In those approaches, position information is stored on a hierarchy of regions (squares) and lookup requests are routed along a list of pointers mapping the logical hierachy to the physical network topology. The disadvantage of such an approach is that – similar to cluster based service discovery [124, 125] – maintaining the hierarchical structure becomes an expensive task under node mobility.

11.8 Summary

In this chapter we have presented *MAND*, a novel infrastructure for implementing networking services in mobile ad hoc networks. *MAND* builds upon the observation that many networking services can be seen as mapping functions. Accordingly, *MAND* implements an efficient mechanism for storing, distributing, and searching

11.8. SUMMARY

tuples (key/value) pairs in MANETs. $MAND$ differs from existing work in that it combines three properties in a unique way: a) parameterized modes of operation to address trade-offs such as update performance vs consistency or lookup performance; b) efficient integration with the underlying routing to minimize overhead and optimize the service access time for network services using $MAND$; c) independence of the type of the underlying routing protocol. Using these ideas, we have implemented $MAND$ over several routing protocols and deployed it on a test-bed consisting of 30 wireless nodes. By extensive experimentation we have shown that $MAND$ performs very well in different settings and different modes of operation. To show the potential and prove the feasibility of $MAND$, we have implemented several networking services that make MANETs a seamless extension of infrastructure based networks.

Chapter 12

Service Discovery and MANET-Internet Integration

As a first use case of *MAND*, we present a flexible solution for the MANET-Internet integration. Our framework allows nodes within the ad hoc network to dynamically discover Internet gateways and establish IP connections to and from the Internet. The framework consists of three components: MANET-SLP, an SLP [133] compatible system for service discovery in ad hoc networks built on top of *MAND*, a *gatewayprovider* providing Internet gateway functionality if Internet connectivity was detected, and a *connectionprovider* connecting to the gateway if one is found. The discovery of Internet gateways is done using MANET-SLP. The connection between *connectionprovider* and *gatewayprovider* consists of a layer two VPN tunnel. At the gateway, the VPN tunnel interface is bridged towards the Internet. Once the client is attached to a gateway using VPN, it will configure its virtual VPN interface using DHCP to gain Internet access. An illustration of a scenario with a gateway and a node in MANET connected to it is shown in Figure 12.1. In the following sections we will describe each of the components (MANET-SLP, gatewayprovider, connectionprovider) in more detail. But first, we give a short overview of the Service Location Protocol as it is defined in RFC 2608 [133].

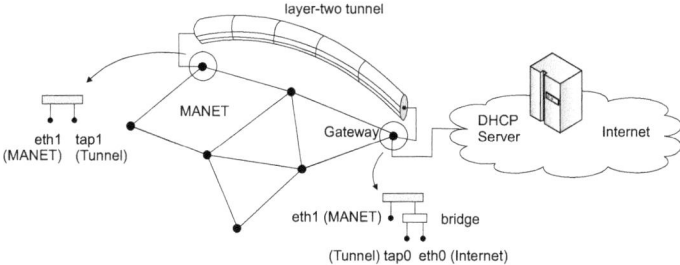

Figure 12.1: Network state after the *Connection Provider* has successfully configured a node for Internet access

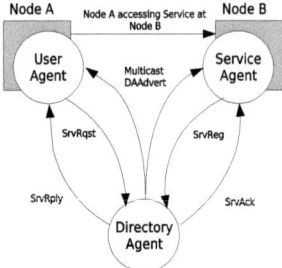

Figure 12.2: SLP Architecture

12.1 The Service Location Protocol (SLP)

The Service Location Protocol (SLP) is a network protocol that allows devices to find services in a local area network without prior configuration. SLP has been designed to scale from small, unmanaged networks to large enterprise networks.

12.1.1 SLP roles

SLP has three different roles for devices. A device can also have two or all three roles at the same time.

- User Agents (UA) are devices that search for services

- Service Agents (SA) are devices that announce services

- Directory Agents (DA) are devices that cache services. They are used in larger networks to reduce the amount of traffic and allow SLP to scale. The existence of DAs in a network is optional, but if a DA is present, UAs and SAs are required to use it instead of communicating directly.

12.1.2 SLP Service

Each service must have a URL that is used to locate the service. Additionally it may have an unlimited number of name/value pairs, called attributes. The URL of a gateway service could look as follows:

`service:printer:lpr://myprinter/myqueue (format=postscript),(color=false)`

This URL describes a queue called "myqueue" on a printer with the host name "myprinter". The protocol used by the printer is LPR. Note that a special URL scheme "service:" is used by the printer. "service:" URLs are not required: any URL scheme can be used, but they allow you to search for all services of the same type (e.g. all printers) regardless of the protocol that they use. The first three components of the "service:" URL type ("service:printer:lpr") are also called service type. The first two components ("service:printer") are called abstract service type.

The attributes of the printer are specified in the last part of the service description ("(format=postscript),(color=false)"). The example uses the standard syntax for attributes in SLP.

12.1.3 SLP Queries

SLP allows several query types to locate services and obtain information about them. It can search for all services with the same service type or abstract service type. The query can be combined with a query for attributes, using LDAP's query language.

12.1.4 Protocol

SLP is a packet-oriented protocol. Most packets are transmitted using UDP, but TCP can also be used for the transmission of longer packets. Because of the potential unreliability of UDP, SLP repeats all multicasts several times in increasing intervals until an answer has been received. All devices are required to listen on port 427 for UDP packets, SAs and DAs should also listen for TCP on the same port. Multicasting is used extensively by SLP, especially by devices that join a network and need to find other devices. Figure 12.2 illustrates the SLP Architecture in a situation with nodes, A and B, where node A is requesting a service from node B.

12.2 MANET-SLP

MANET-SLP is a fully distributed service discovery platform for MANETs. It provides a regular SLP interface for service registration and lookup. Services can be registered using the SLP *REGISTER* interface and looked up using the SLP *LOOKUP* interface. Logically, MANET-SLP resides on top of MAND. However, rather than storing service information into the MAND system, MANET-SLP uses MAND to distribute SLP events across the network, similar as shown in Figure 11.3. Suppose MANET-SLP receives a request to register the printer service we have described previously. MANET-SLP will cache the service information locally and additionally it will insert the following tuple into the MAND system:

```
tuple{
    key=slp,
    value=service:printer:lpr://myprinter/myqueue
        (format=postscript)(color=false)
        service-lifetime=<service-lifetime>,
    lifetime=0,
    scope=<scope>
}
```

All tuples exchanged between MANET-SLP and MAND belong to the *slp* namespace and thus have a common key. Every tuple is issued with a *lifetime* of zero. As a consequence, the tuple is not stored in MAND, but is triggering events to all current subscriptions for **key**=slp. Every MANET-SLP instance running on a node maintains such a subscription with the underlying MAND system. The subscription has previously been issued and is periodically renewed using a MAND lookup call for the key *slp*. The tuple request passed to the *MAND lookup* interface may look as follows:

```
request{
    key=slp,
    lifetime=100,
    scope=0
}
```

The zero *scope* of the tuple request makes sure the request is not propagated into the network. The *lifetime* of 100 makes sure that the request stays in the *MAND* system for 100 seconds. During this period, *MAND* delivers all tuples with key "slp" to *MANET-SLP*. Before the lifetime of the request expires, *MANET-SLP* will re-insert the request to update the subscription.

For every event MANET-SLP receives, MANET-SLP will store the service information given by tuple's value in its local cache. How long a service information is cached locally is determined by the service lifetime which is an optional parameter of the SLP interface. If it is specified upon calling *SLP REGISTER*, the *service-lifetime* will be encoded into the tuple value during the MAND insert call, which makes the *service-lifetime* available to remotely cached services. SLP services registered with MANET-SLP may have a special attribute *scope*. The value of the *scope* attribute is directly mapped to the scope of the corresponding MAND tuple, thus determining the visibility of the service in the ad hoc network. There are default values for both *service-lifetime* and *scope* that can be defined upon system startup.

SLP Lookukp queries received by MANET-SLP will be answered just by consulting the local cache. There might be multiple services matching a given request. In MANET-SLP, all the service related information is exchanged pro-actively using MAND, therefore no additional traffic is imposed to the network.

12.3 MANET-Internet integration

The MANET-SLP service can be leveraged to build a complete MANET-Internet integration.

12.3.1 Gateway Provider

A *Gateway Provider* is a process that can set up a node to become a *gateway* in case the node has Internet connection, and removes the *gateway* functionality in case the Internet connection is lost. *Gateway Provider* processes are started on nodes willing to act as gateways. A *gateway* is a node that is directly connected to the Internet and configured to provide Internet access to all the nodes within the MANET. Typically, these are nodes with multiple interfaces, since one interface is configured for communication with the MANET, and another interface is dynamically attached to the Internet. How a *Gateway Provider* works is shown in Figure 12.3. Once started, the *Gateway Provider* process keeps waiting for an Internet connection to become available (Figure 12.3, step 1). Detecting whether a node has Internet connection or not is done using a special *InternetDetectionAPI*. The idea is to exploit system support to efficiently detect a possible Internet connection. Our current implementation makes

12.3. MANET-INTERNET INTEGRATION

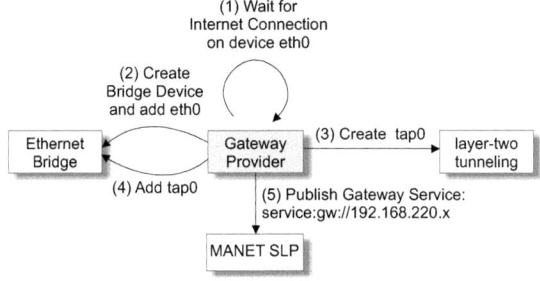

Figure 12.3: Gateway Provider

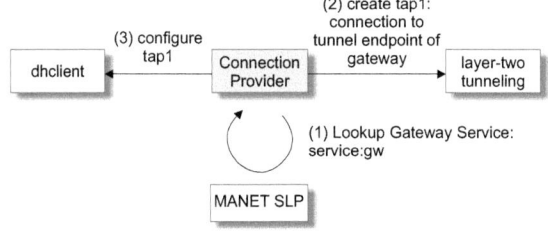

Figure 12.4: Connection Provider

use of the operating system routing table. If an Internet connection has successfully been detected, say on interface *eth0*, the *Gateway Provider* process then creates a *bridge* device on that node and immediately adds interface *eth0* to the *bridge* (step 2). In a following step, a *layer-two tunnelling* device *tap0* is created (step 3) and also added to the *bridge* (step 4). The device *tap0* allows any node within the MANET to set up a *layer-two tunnel* connection to the gateway node. Since the tunnel device *tap0* is part of the *bridge*, traffic received on *tap0* is directly forwarded to the Internet via interface *eth0*. To provide MANET nodes with such gateway functionality, a *Gateway Provider* must however first register itself as a *gateway* service using the underlying *MANET SLP* service (step 5). The SLP service description may look as follows:

```
service:gateway:manet://10.4.0.17
```

Once a gateway service is registered, any node within the MANET may look up the gateway's location and connect to it. If the Internet connection is lost the service will be de-registered or will timeout, and both the tunnel endpoint *tap0* and the *bridge* will be removed.

12.3.2 Connectionprovider

A *Connection Provider* is a process that sets up an *Internet connection* if a gateway can be found. *Connection Provider* processes are started on nodes who want an Internet connection. How a *Connection Provider* works is shown in Figure 12.4. Once started, the *Connection Provider* process periodically searches for a gateway service by performing an SLP lookup request (Figure 12.4, step 1). If a gateway service can be found[1], a *layer-two tunnel* connection to the gateway is established (step 2). To finally configure the node for Internet access, a DHCP request is triggered on that newly established *tap1* interface (step 3). Since the *tap0* interface at the gateway node is bridged towards the Internet, the DHCP request will eventually be answered by the DHCP server that is reachable from the gateway node. The mechanism of IP configuration is encapsulated in an *IPConfiguration* module with a well defined interface. This allows the *Connection Provider* to easily adapt to other ways of IP configuration such as, e.g., IPv6 auto-addressing or MobileIP. After the IP configuration on the *tap1* device is done, the corresponding node is not only able to communicate with any node in the Internet, but nodes from the Internet may also transparently connect to that node within the MANET. A more detailed perspective on how components such as *bridge* and *layer-two tunnel* interfaces interact with each other is given in Figure 12.1.

The proposed mechanism differs from previous work for MANET-Internet connectivity in that it combines both a dynamic approach (through the use of a Gateway-Provider and a Connection-Provider) with a routing independent approach (through layer-two tunneling) while still being message efficient (due to MANET SLP).

12.4 Performance

One important question is how quickly after a new gateway has been created can it be discovered by other nodes in the network and how fast is the service lookup. Since MANET-SLP acts pro-actively, the service lookup time corresponds to the time of a local cache lookup which is negligible. However, it takes some time for new service information to become available to other nodes. This time corresponds to the tuple propagation time that is discussed in sections 11.6.1 and 11.5.3.

To see whether tunnel maintenance affects the performance of the gateway, we have measured *dial-to-ring* delays for various setups with up to 1000 additional (idle) tunnel connections, without observing any recognizable slowdown.

We have also studied the packet overhead caused by the fact that the gateway is accessed through a layer 2 tunnel which wraps the packet and adds its own Ethernet and IP headers. The Ethernet header uses 14 bytes and the IP header 20 bytes. *openvpn*, used as tunneling application in our setting, sends its packets using UDP which adds another 8 bytes. This results in an overall overhead of 42 bytes per packet. Compared with the typical MTU of 1500 bytes per packet in Ethernet, an overhead of 42 bytes is almost irrelevant. For voice data, however, the audio data contained in a UDP packet is typically in the range of 160-172 bytes. A voice data packet on the wire without the tunneling overhead would therefore have a size of 214 bytes. With

---

[1]Currently, if multiple gateway services are found, the list of all gateways is passed to the tunnelling component

the tunnel header, the size of each packet increases to 256 bytes. This is an overhead of 20%. If only a few nodes in the network communicate through the gateway using tunneling this will not have a big effect. If the network is big and most of the users communicate through the gateway to the Internet, an overhead of 20% may decrease the available capacity.

12.5 Related Work

12.5.1 Service Discovery

Some conventionally adopted protocols, including Sun's Jini [134], the service directory service UDDI [135], and IBM's Salutation [136] rely on a central directory to register and discover services. Centralized architectures are inappropriate for MANETs because of the dynamic nature of such networks. Moreover, centralized directories easily suffer from routing bottlenecks.

Some distributed approaches, including Universal Plug and Play (UPnP) [137], Service Location Protocol (SLP) [133] of the IETF, and Bluetooth's Service Discovery (SDP) [138], have been developed to reduce the overload of the centralized directory. However, those protocols are based on a network-wide multicast tree used for periodic service advertisements. Multicast, however, is known to perform poorly in MANETs.

There are approaches specifically targeted at ad hoc networks. The distributed service discovery protocol (DSDP) [139] has been developed to register the information of available services to a virtual backbone, and is based on clustering. A similar approach has been proposed by [124]. However, electing and maintaining specialized backbone nodes is difficult and expensive due to the mobility of MANETs. In fact, it has been shown [125] that in terms of transmitted messages, a fully distributed service discovery outperforms the per-cluster approach, especially in mobile scenarios. To cope with mobile scenarios it has been suggested to integrate service discovery with routing in MANETs. In [140], the authors propose to add extensions to the On-Demand Multicast Routing Protocol (ODMRP) [141] to advertise service information. A similar approach was presented in [142, 143] based on the Optimized Link State Routing Protocol and in [144] for the Ad Hoc On Demand Routing Protocol (AODV). It is true that by leveraging functionality of MANET routing protocols, service discovery can be improved, especially in terms of mobility support. However, we believe that a tight integration is impractical since it enforces service discovery clients to use a specific routing protocol. This is a loss of freedom because there are plenty of different approaches for routing in MANETs, each of which has its own advantages and disadvantages in certain situations.

12.5.2 MANET-Internet integration

Dynamic gateway discovery has been combined with MANET routing [145]. While this might be efficient, the solution is tailored to a specific routing protocol and not generalizable. Once a gateway is discovered, the challenge is to find ways to route SIP and voice traffic through the gateway. In [146, 147, 148], specific functionality of the given MANET routing protocol is exploited to forward data to the Internet. However, these solutions lack generality since they are routing specific. A routing independent approach, where IP tunnelling is used between the end nodes and the gateway, can be found in [149], but their work requires gateway nodes to be located according to a fixed hierarchy. A more flexible design without topology assumptions is described in [150, 151]. IPv6 auto-addressing for IP endpoint configuration has been proposed in [145, 152], which is also not generic considering that IPv6 is not yet widely deployed. In order to provide a fixed IP endpoint to the outer world in case the gateway changes, Mobile IP has been suggested [150, 151, 146, 147, 148]. Alternatively NAT is used for addressing. In the case of VoIP, a NAT approach would require some additional mechanisms like STUN [153] to allow peers located in the Internet to establish calls to nodes located within the MANET.

12.6 Summary

In this chapter we have presented MANET-SLP, a distributed and SLP compatible system for service discovery in MANETs. Based on MANET-SLP we have described a framework for MANET-Internet integration. The framework allows nodes in the MANET to communicate to and from the Internet as soon as one node in the ad hoc network has Internet connectivity. In the next two chapters, we will make use of the MANET-Internet framework to allow DNS resolution and SIP routing to happen seamlessly accross the boundaries of MANET and Internet.

Chapter 13
DOPS: Domain Name and Presence Service

Another example of a standard networking service, besides SLP, is DNS. In this chapter, we describe a Domain Name and Presence Service (*DOPS*) service, tailored to ad hoc networks. *DOPS* provides a standard DNS interface but its implementation is in the form of a fully decentralized middleware platform. *DOPS* uses MAND as a directory service and the MANET-Internet integration framework presented in the last chapter to integrate seamlessly with the Internet. Moreover, *DOPS* naturally interlinks the ability to resolve a certain hostname with the presence of the corresponding node in the network.

13.1 Standard DNS

Conceptually, DNS defines a tree where a node in the tree is identified by a particular domain name. Each node is associated with a set of resource information, so called resource records (RRs) [154]. Most importantly, resource records hold data like the IP address of a node. There are many different types of resource records, for instance one to store an IPv4 address or one to store an IPv6 address. A DNS query names the domain name of interest and describes the type of resource information that is desired. Nameservers hold information about some part of the domain tree's structure and its associated resource records. The essential task of a name server is to answer queries about the resource records the nameserver is associated with.

13.2 DOPS in the MANET

DOPS is a distributed domain name system running as a user space process on each node in the ad hoc network[1]. Each *DOPS* instance stores the resource records associated with the host node's own domain name in *MAND*, thereby providing full DNS access to all nodes in the network. To make use of *DOPS* a user has to just configure localhost to be the primary DNS contact address.

---
[1]*DOPS* may only run on a subset of nodes in a MANET; those that are interested to participate in a MANET-wide DNS system

13.2.1 Mapping Resource Records to MAND Tuples

DNS resource records contain four important fields: The **name** specifies the domain name the RR is associated with. The **type** specifies the type of the RR (e.g., host address type or a well known service description type). The **class** field identifies a protocol family (e.g., IN for Internet System or CH for Chaos System). The **ttl** describes how long a RR can be cached before it should be discarded. Finally, **rddata** is the **type** dependent data which describes the resource (e.g., the actual host address). *DOPS* maps the structure of a RR to a MAND tuple as follows.

```
tuple{
    key=dns:<name>:<type>:<class>,
    value=<rddata>,
    lifetime=<ttl>,
    scope=64
}
```

The name, the type and the class of the RR is used as a key of the tuple, the RRs rddata field (an IP address) forms the value. We use the RRs ttl to specify the lifetime of the tuple. This makes sure that there are no stale RRs for nodes which are not present anymore in the network. In order to keep remote entries persistent for the time the corresponding node is active, *DOPS* will periodically re-distribute the tuple. The period in which a *DOPS* instance advertises the node's own RR (advertisement period) can be configured at startup. There is obviously a relationship between the advertisement period and a RR *ttl* value. *DOPS* sets the *ttl* of the node's own RR to a multiple of the advertisement period. The multiplication factor can also be configured at startup. Typical values for the multiplication factor are 2 or 3. The advertisement period and the multiplication factor allow to control how fast *DOPS* reacts to network changes. In the best case, any change in the presence status of a node (joining the network, leaving the network) is immediately communicated across the network. Obviously, there is a trade-off between the agility of the system and the load DOPS imposes to the MAND system. We are going to study this effect in more detail in Section 13.4.

13.2.2 Queries

A standard DNS query specifies a target domain name, query type, and query class and asks for resource records that match. In *DOPS*, upon receiving a query, the corresponding domain name entry is looked up in the local *MAND* instance by issuing a *MAND* lookup request.

```
request{
    key=dns:<name>:<type>:<class>,
    lifetime=0,
    scope=0
}
```

The key used when issuing a MAND lookup request is composed of the name, the type and the class fields of the query, similar as shown in the previous section for

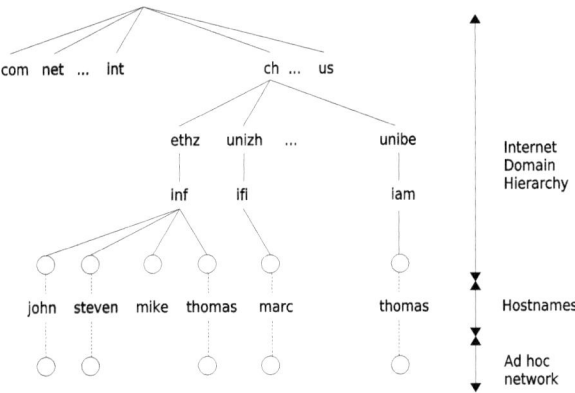

Figure 13.1: MANET-Internet integration

the insertion of a RR. The zero *scope* makes sure the request is not propagated into the network. The *lifetime* of zero makes sure we get a reply immediately. If the MAND lookup request is successful, *DOPS* responds with a DNS reply containing the resource record (including the IP address) for this entry. If no valid tuple can be found for the requested domain name, the *DOPS* responds with a DNS error code. Since each *DOPS* instance has access to all the resource records of all active nodes in the MANET, a DNS lookup request can be handled locally, which is optimal.

13.2.3 Distributing RRs

Each *DOPS* instance running on a node is responsible for at least two RRs associated with that node. For instance, a node with the hostname *node1* in the domain *ethz.ch* would have RRs for the names *node1* and *node1.ethz.ch* assigned. Those two RR support queries for the fully qualified name, but also for the hostname only.

The Internet uses the `in-addr.arpa` domain to support Internet address to host mapping. For instance, if resolver wanted to find the host name corresponding to a host address `10.4.0.6` it would pursue a query of the form **type**=PTR, **class**=IN, **name**=`6.0.0.10.IN-ADDR.ARPA`. Name servers can support `in-addr.arpa` queries through exhaustive searches of their databases, but this becomes impractical as the size of the database increases. An alternative approach, is to invert the internal store according to the search key. *DOPS* does so by providing a MAND tuple for the inverted resource record of the node. This is additional to the two RRs already provided. Thus, in total one *DOPS* instance provides 3 RR to the underlying *MAND* system.

13.3 Seamless integration with the Internet

The ultimate vision of a DNS infrastructure that works in both isolated and Internet connected MANETs is that nodes can use their fully qualified domain name (FQDN) transparently within the MANET as well as in the Internet. In other words, the only change should be the scope of the corresponding domain name which will vary depending on whether the MANET is currently connected to the Internet or not. For instance, assume a node in the Internet with domain name *node1.ethz.ch*. Given that this node is temporarily detached from the Internet but attached a MANET, we would like communication with other nodes in the MANET based on the node's Internet domain name (*node1.ethz.ch*) to be possible without applications to take notice. On the other hand, communication to and from the Internet should be possible as soon as the node gains Internet connectivity again.

13.3.1 Example

Imagine a scenario as illustrated in Figure 13.1. A research project involves several research groups from different universities in Switzerland. Occasionally, the members of the various groups will meet to discuss the latest state of the project. During those meeting the people involved might want to exchange certain data related to the discussion they have. However, rather than building a fixed network to exchange data a MANET could be used. It would be important to have DNS infrastructure that provides the very same view on the network as if the nodes would be attached to the Internet. This would allow applications that exchange data to run in the MANET as if they would run in the Internet.

13.3.2 Query processing in Internet-connected MANETs

DOPS does not provide domain name resolution for nodes outside the ad hoc network. Rather, it forwards those DNS requests to the Internet. To implement the vision of transparent and seamless DNS connectivity, the *DOPS* system needs to decide whether a node with a given domain name can be reached over the MANET or over the Internet. There is no point in making such a decision if the *DOPS* instance receiving the DNS request runs on a node without Internet access. If, however, a node is attached to both MANET and Internet[2], things become more complicated. We refer to a node in this state as a *hybrid* node. One way to cope with hybrid nodes would be to first forward the query to the Internet and trying resolve a name in the MANET only if the Internet query has previously failed. This creates an overhead in the DNS response time while answering domain names that are currently only reachable within the MANET. To cover the worst case, a *timeout* is needed to decide whether a given name can be resolved in the Internet or not. And even if a DNS query sent to the Internet is successfully answered, it still does not tell anything about whether the given node is currently reachable through the Internet. Another possibility, the one implemented in *DOPS*, is to query the MANET first and only in case of a failure trying to resolve a name in the Internet. The advantage is that querying the MANET in *DOPS* does not create any latency overhead since this corresponds just to a local lookup. One can argue that for nodes attached to both Internet and MANET it would make more sense to respond with the IP address that belongs to the Internet since communication over the Internet typically is faster. There are two answers to this question. First, a MANET composed of nodes that are

---

[2]DOPS detects an Internet connection using the InternetDetectionAPI described in chapter 12

DNS Configuration	Query Type	Available Network Connections		
		Ad Hoc	Ad Hoc/Internet	Internet
DOPS	Ad Hoc	6.8ms	6.9ms	9.3ms
	Internet	6.8ms	9.8ms	9.9ms
Regular DNS configuration	Internet	N/a	N/a	10.1ms

Table 13.1: Average DNS response times taken from 10'000 measurements

at the same time attached to the Internet is not very common. Optimizing for this case would be suboptimal in many other cases which are far more likely (a MANET composed of nodes not attached to the Internet). The second answer is that we can provide support for nodes that want to prioritize communication through the Internet. In *DOPS* this is done by configuring a separate domain name for usage only in the MANET. A node may want to configure *DOPS* in such a way if the node spends most of its time in a *hybrid* state communicating with other *hybrid* nodes. The separation of the domains gives priority to the communication over the Internet. The disadvantage is that the MANET-Internet integration under such a configuration is no longer transparent to upper layers.

13.4 Performance

We have evaluated *DOPS* both in real and by simulation. The performance metrics we studied are response time, update time and system agility (the ability to react to frequent changes of the network under heavy load).

13.4.1 Response time

First we measured the response time of *DOPS* under different network settings. The measurements were done using 6 notebook computers running Debian 4.0. The laptops had a 2.0Ghz Mobile Pentium 4 and were equipped with an integrated 11Mbit/s IEEE 802.11a wireless network interface card. During some of the measurements, one laptop was additionally attached to the Internet using a Gigabit Ethernet network card. Because it would be difficult to find a spatial separation of the notebooks which would have required multihop communication between them, an artificial separation using packet filter rules was used: They were only allowed to communicate with their direct neighbors, all other traffic was dropped (by the default policy). All results are averaged over a set of 10'000 tests.

The test settings differ in terms of whether the node submitting a DNS query is attached to the Internet access, or the ad hoc network, or to both the Internet and the ad hoc network. Additionally, we distinguish the case where the domain name to be resolved concerns a node in the ad hoc network, or a node in the Internet. The former is called a query of type *Ad Hoc*, the latter a query of type *Internet*.

From table 13.1 we see that resolving a domain name for a host in the ad hoc network, when being attached to the ad hoc network, takes $6.8ms$ on average. The response stays approximately the same (6.7ms) if the node additionally is attached to the Internet. This is because in any case, *DOPS* first checks whether a given domain name can be resolved in the ad hoc network[3]. If a node tries to resolve a domain name for a host that lies in the Internet, while the node itself is only attached to the ad hoc network, it takes as well around 6ms until a negative response is received. If, however, the node additionally is attached to the Internet, the domain name will successfully be resolved with a response time of 9.8ms on average. This is because *DOPS* acts as a proxy and forwards the DNS query to the nameserver in the Internet. We made some measurements to see what the overhead is of using *DOPS* for communicating with the Internet. We see that even if a node is solely attached to the Internet, and detached from the ad hoc network, the average response time stays around 9.9ms, which is roughly the same as in the case where the node was additionally attached to the ad hoc network. Therefore, the overhead of trying to resolve a domain name in the ad hoc network first is negligible. This is because looking up a domain name in the ad hoc network is only a local operation. We have also compared the response time for querying Internet host using *DOPS* versus sending DNS queries directly to the Internet domain name server. From the Table 13.1 we see that the response time without using *DOPS* as a proxy stays approximately the same. Therefore, using *DOPS* will not cause any overhead when communicating with the Internet, but seamlessly integrates with the domain name system in the Internet.

13.4.2 Update time

The update time is the time it takes for *DOPS* to distribute a new resource record for a given domain name. This is important since it determines how fast the system reacts when a new node joins the network or an existing node leaves the network. Since *DOPS* uses *MAND* to distribute the RRs across the network, the update time corresponds to the tuple propagation shown in Figure 11.6 in chapter 11. According to those figures, it takes around 0.8 seconds on average until its domain name information is available to every node in the network, considering a network with 30 nodes and a diameter of 5. Figure 11.9 further indicate that mobility has a positive effect on the update time.

13.4.3 Presence Detection

We have used $ns - 2$ to explore the behavior of *DOPS* in a larger scale. Our configurations span from a basic configuration of 50 nodes deployed uniformly in an area of $1500m \times 300m$, up to configurations with 100, 200, 500, and 1000 nodes spread across planes appropriately enlarged to maintain the aspect ratio and node density of the basic configuration. We carried out our experiments both over AODV and OLSR using the same implementation we used already in chapter 11 to evaluate the *MAND* system.

In the first experiment we study the impact *DOPS* has on *MAND* when distributing a large set of RRs/tuples. Remember that the *MAND routing handler* stores all tuples while they are waiting to be piggybacked onto an intercepted routing message. Let's assume the datastructure to store those tuples is called *SendQueue*. In the

---
[3]Note that this behavior can be changed in *DOPS* by configuring a special domain name for the node to be used in the MANET

13.4. PERFORMANCE

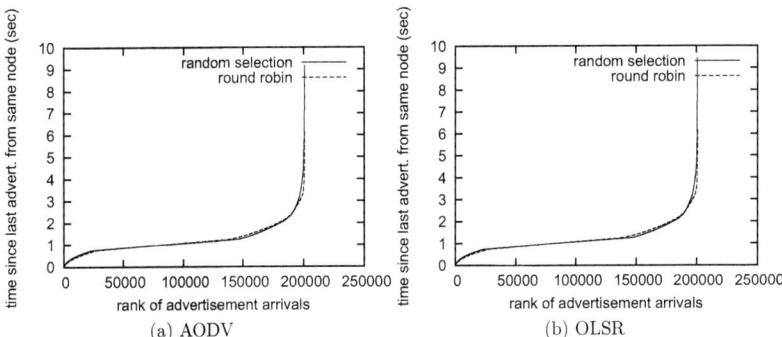

Figure 13.2: The span of intervals between consecutive advertisements of node Q arriving at node P, for all pairs of nodes P and Q. The *random selection* and *round robin* policies are shown.

worst case, the *SendQueue* holds a RR/tuple for every node in the network. If a routing message is intercepted, there might not be enough space in left in the message to piggyback all the RRs in the *SendQueue*. One option is to pick RRs in a round robin fashion. Another approach is to always pick a randomly chosen subset of RRs. Given the inherent redundancy of dissemination paths in ad hoc networks, the second option (choosing a random subset) is more appealing. In the simulation, we consider both alternatives. First, a node piggybacks a *random subset* of the RRs it currently holds in the *SendQueue*. Second, a node picks RR from its *SendQueue* in a round-robin fashion.

We ran experiments with 50 nodes, each node injecting a his own RR in every outgoing HELLO packet (once per second). Nodes additionally forwarded up to 20 RR advertisements of other nodes in each packet. We let the experiments (for both AODV and OLSR) run until 4'000 RRs were injected in total. All RR advertisements reached all nodes, resulting in a total of 196'000 advertisement arrivals. For each RR of node $n1$ arriving at node $n2$, we logged the time elapsed since $n2$ last saw a RR of $n1$, for all pairs of nodes. Essentially, we recorded the intervals at which a node gets to hear about the existence of some other node. Figures 13.2a and 13.2b plot these intervals, ordered based on their durations. We observe that RR advertisements arrive mostly every one or two seconds, with a few exceptions. Given that the advertisements are issued once per second, we infer that the vast majority of the network stays closely up to date with a node's existence. Note that the difference between the *random selection* and *round robin* policies is hard to notice in this graph, but will be shown subsequently.

13.4.4 Adjusting the advertisement expirations (*ttl*)

As mentioned in Section 13.2.1, each node applies an expiration time (ttl) to each RR received. If the last RR advertisement of node $n1$ in node $n2$'s *MAND* system has expired, $n2$ considers $n1$ to be dead. Clearly, a high *ttl* value, at least as high as the maximum expected interval between consecutive advertisements, will prevent a

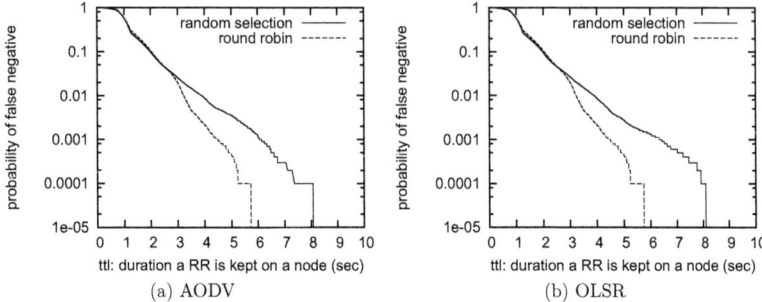

Figure 13.3: The probability for a node to falsely consider another node dead (false negative), as a function of the *ttl*. Comparison between AODV and OLSR, and between *random selection* and *round robin*.

node from assuming another node dead due to some delay in its RR advertisement. On the other hand, with a very high *ttl*, dead nodes are remembered long after their death and considered alive. Therefore, setting an appropriate value for *ttl* is a crucial tuning parameter of *DOPS*.

Figure 13.3 shows the probability of a node falsely considering another node to be dead, as a function of the *ttl*, for a network of 50 nodes. The main observation is that, irrespectively of the policy used, *random selection* or *round robin*, a *ttl* of less than 10 seconds can allow nodes infer other nodes' death with accuracy. A second observation is that AODV and OLSR have the same behavior with respect to *DOPS* and RR advertisement updating.

Interestingly, *round robin* performs better than *random selection*. In *random selection*, some consecutive advertisements are delivered up to 8 seconds apart, while in *round robin* the maximum interval observed between consecutive advertisements is closer to 6 seconds. It is not hard to see why. *Random selection* picks the RR advertisements to forward at random. This is asymptotically fair to all advertisements, but may lead in some temporal imbalance between the number a specific RR has been forwarded in some neighborhood of the network. *Round robin*, on the other hand, provides a much more predictable schedule in forwarding RRs, diminishing the the worst case.

Finally, Figure 13.4 presents similar graphs for larger networks, or 100, 200, 500, and 1000 nodes. To avoid duplication of equivalent graphs, only AODV graphs are shown, but OLSR have been double-checked to appear the same. As expected, larger networks impose higher delays in RR propagation, therefore a higher *ttl* value is needed. For instance, for the case of 1000-node networks, nodes should be keeping advertisement for at least 60 seconds to avoid falsely regarding other nodes as dead. However, if false negatives are not critical to the application using *DOPS*, a lower *ttl* value can be chosen, which will also allow the faster detection of dead nodes at the cost of some false accusations. In our example of the 100-node network, a *ttl* of 20 seconds would allow nodes to detect dead nodes three times faster, at the cost that 10% of the cases that a node is considered dead would be false.

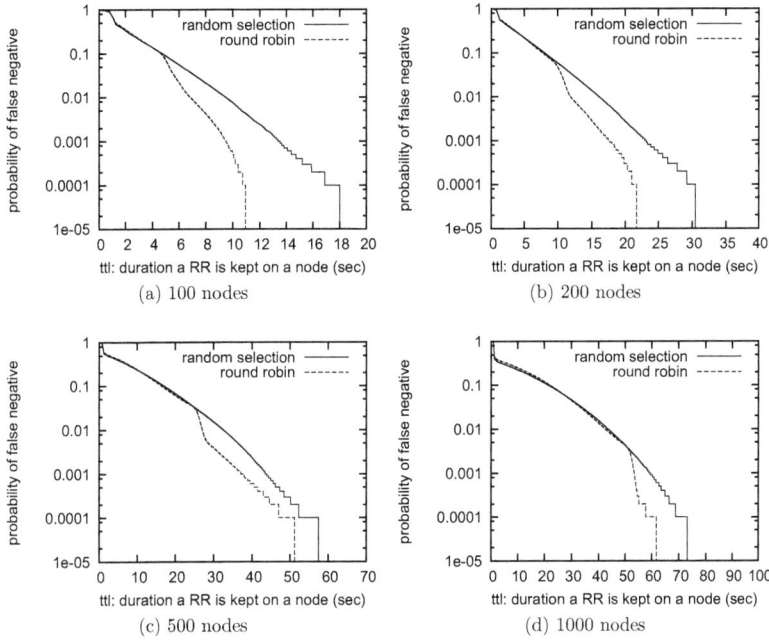

Figure 13.4: The probability for a node to falsely consider another node dead (false negative), as a function of the *ttl*. Simulations of larger networks

13.5 Related Work

A *natural* way of performing domain name resolution is to use a decentralized approach in which a node of the MANET replies to a broadcasted name request for which it is the target [155, 156, 157]. One problem with this approach is that it uses an excessive amount of messages due the broadcasts triggered every time a domain name needs to be resolved. Broadcasting DNS queries is a particular waste of resources if the node matching the requested domain name is not present in the network. Another problem is that resolving domain names on demand prolongs the overall communication setup time (mainly consisting of the route establishment time). In [158, 159], the overall amount of messages as well as the communication setup time is lowered by integrating route discovery and domain name resolution. Another option is to use multicast routing to propagate domain name mappings [160, 161]. However, all of those approaches are bound to one particular routing protocol, e.g., AODV [93] or DSR [122].

A number of solutions exist for providing DNS across the MANET-Internet boundaries. In [159, 162], the authors propose a special domain *net.lunar* to be configured at startup. An integrated DHCP server then appends the pre-configured domain to all nodes in the MANETs. In this way, the DNS system knows exactly which domain it has to query the Internet for and which not. However, the method requires applications to communicate using the newly defined domain names. Moreover, two nodes respectively named *john.domain1.net* and *john.domain2.com* will both end up being identified inside the MANET as *john.net.lunar*.

Another approach, discussed in [125], is based on specialized *gateway* nodes which forward DNS requests to the Internet and send back the replies. This approach, however, fails if the requested domain can be resolved in the Internet but the node is currently detached from the Internet but available in the MANET. As a solution, the authors propose to *ping* the node before returning the resolved address, but ping requests are often blocked by firewalls. Alternatively, a DNS system could first try to resolve the domain name in the MANET and try the Internet afterwards, but this causes a huge overhead in response time when resolving domain names of nodes which are actively attached to the Internet.

13.6 Summary

In this chapter we have presented *DOPS*, a DNS and presence service for MANETs. *DOPS* is fully decentralized and at the same time seamlessly attaches to the Internet domain name system. *DOPS* uses the *MAND* system for the distribution of DNS resource records. In this way, *DOPS* provides a novel but natural mechanism to combine presence signaling and domain resolution with zero message overhead. We have shown that *DOPS* quickly reacts to network changes and provides a fast DNS response time.

Chapter 14

SIPHoc: Distributed Session Initiation in MANETs

SIP is an important protocol and network service used by many applications in the Internet such as VoIP, video or chat. In this chapter, we present SIPHoc, a SIP compatible middleware for session setup and management in MANETs. SIPHoc provides the same interface as the SIP standard but its implementation is tailored to the decentralized and dynamic nature of MANETs. Similar as *DOPS* for DNS, SIPHoc provides SIP services beyond the MANET-Internet boundaries. We are testing SIPHoc using standard SIP-based VoIP applications to set up a VoIP communication between nodes in the MANET and occasionally also nodes in the Internet.

14.1 SIP Overview

SIP works by building an overlay network on top of a regular IP network using a set of SIP entities such as proxies and registrar servers. A *SIP proxy* is an intermediary entity primarily in charge of routing: its job is to ensure that a request is sent to another entity 'closer' to the targeted user. A *SIP registrar* is used by SIP applications to register their current location. A *SIP location service* is used by registrars to store user location information and by SIP proxies to query user location information.

A SIP application first registers with the system by communicating its *SIP user name* and its current location to a *registrar* (Figure 14.1a). The registrar to be used is either determined using DNS or it is statically configured (through the so called *outbound-proxy*). Registrars and proxies are logical entities and it is not uncommon to have them co-located on the same node. When the registrar hears from a node, it builds a *binding*, i.e., an association between a SIP user name and the corresponding contact address (typically an IP address or resolvable name).

To establish a session with another user whose current location is unknown, a *SIP INVITE message* is sent to the proxy/registrar (step 1, Figure 14.1b). The proxy responds with a *100 Trying message* (step 2). The 100 Trying response indicates that the INVITE message has been received and that the proxy is trying to route the INVITE message to the final destination. Since the outbound proxy/registrar does not know the location of user B, it uses a DNS server to locate the proxy of the destination node and forwards the INVITE message accordingly (steps 3 to 6). The receiving proxy/registrar uses the previously registered binding information of

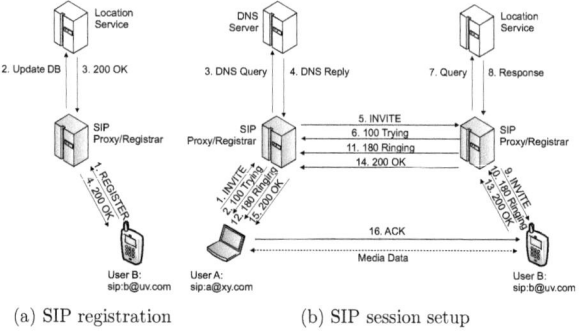

(a) SIP registration   (b) SIP session setup

Figure 14.1: Basic SIP mechanism

the user to locate the destination (steps 7 and 8) and finally delivers the INVITE message to the intended recipient (step 9). The recipient responds with a *180 Ringing message*, which is routed back through the two proxies in the reverse direction (step 10 to 12). If user B decides to establish the session with user A, it responds with a *200 OK* (step 13 to 15). Finally, user A sends an ACK message to confirm the reception of the 200 OK message (step 16). At this stage, the two users have learned each others' contact address through the INVITE/200 OK messages and from then on they communicate directly, bypassing the two proxies.

14.2 SIPHoc in MANETs

The basic principle of SIPHoc resembles the one of DOPS in that we run an instance of a modified SIP proxy and registrar (SIPHoc proxy) on each node in the network[1].

14.2.1 SIPHoc proxy

A standard SIP proxy/registrar accepts SIP registrations of a collection of users from certain domains. A SIPHoc proxy typically only accepts SIP registrations from users (applications) on that particular device. In addition to storing these registrations in its local location service table, each SIPHoc proxy uses *MAND* to advertise itself as the contact address for these registered users (e.g., as the outbound proxy for those users). If a SIPHoc proxy receives an INVITE message and cannot find the target in its local location service table, it consults *MAND* and forwards the INVITE message to the proxy which was advertised as outbound proxy for this user (we call this procedure *dynamic outbound proxy selection*).

An alternative approach would have been to implement a distributed storage solution using *MAND* where each proxy knows the direct contact address of all other SIP users in the MANET. Such a solution may look even more appealing but it creates SIP incompatibilities. If each proxy can find out the location of the final destination of the session by itself (using *MAND*), SIP bindings are resolved directly

---
[1] Again, a *SIPHoc proxy* may only run on nodes interested in participating a SIP service

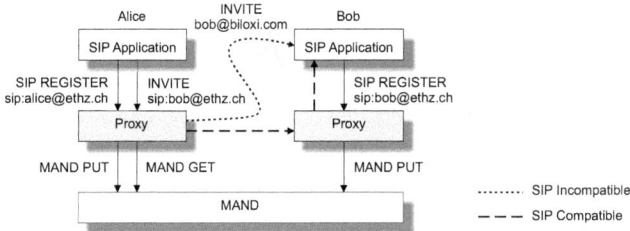

Figure 14.2: Operations of a *SIPHoc* proxy

at the caller's proxy and INVITE messages do not pass the callee's proxy. This violates the standard SIP message flow, where INVITE messages always pass the proxy where the target has registered in the first place. The modified message flow creates problems with session mobility and session tear down.

In contrast to this, SIPHoc complies with the traditional SIP message flow. Moreover, the use of a proxy at each device creates an infrastructure that is fully decentralized by design. Figure 14.2 illustrate both the message flow used by the *SIPHoc* proxy (SIP compatible) as well as the message flow that creates SIP incompatibilities.

14.2.2 Example

How the *SIPHoc proxy* works can be best understood with an example. The example maps the standard SIP message flow shown in Figure 14.1. We assume two users Alice and Bob. The IP addresses of the two machines of Alice and Bob are 192.168.220.1 and 192.168.220.2 respectively. Each user runs a *SIPHoc Proxy* on port 5060 and a SIP application on port 5062. Both machines are in the MANET, within an arbitrary hop-distance from each other. *ProxyA* is the proxy used by Alice and *ProxyB* that of Bob.

To register with SIPHoc, both users send their URI and contact address to their proxies: *sip:alice@ethz.ch* and 192.168.220.1:5062 for Alice; *sip:bob@ethz.ch* and 192.168.220.2:5062 for Bob. The local SIPHoc proxy for each user will then store the corresponding entry in its local location service table. It will also insert a tuple into *MAND* associating his contact address with the given user. The tuple inserted into *MAND* by *ProxyA* on behalf of the registration of Alice may look as follows:

```
tuple{
    key=sip:alice@ethz.ch,
    value=192.168.220.1:5060,
    lifetime=<sip-lifetime>,
    scope=64
}
```

The *lifetime* of the tuple is inherited from the SIP lifetime that was specified in the SIP registration message. Note that the entry in the local location service table differs from the entry sent to *MAND* in that the latter contains the contact address of the proxy (port 5060) rather than the one of the user (port 5062). To establish a SIPHoc session – assume Alice contacts Bob – Alice sends an INVITE message to *sip:bob@ethz.ch* (9). *ProxyA* checks whether the target SIP URI is in the local location service table (10). If that is not case, it uses *MAND* to look up the correct outbound proxy for the SIP URI (11-12). The tuple request issued to *MAND* may look as follows:

```
request{
    key=sip:bob@ethz.ch,
    lifetime=0,
    scope=64
}
```

The *lifetime* of zero makes sure a reply is sent immediately by *MAND*. As we will see later (section 14.3, this is important when connected to the Internet. The *scope* of 64 makes sure that, in case there is not matching tuple available locally, the request is propagated to the network. Any valid tuple that is found in the network will not be considered for the reply that is sent for this request, but it may be considered for the next request that received by *MAND* at a later point in time. Many SIP applications, if they cannot establish a connection on the first attempt, will try for a second time after a timeout. In this case, *MAND* would have pre-fetched the corresponding outbound-proxy information.

Once the outbound proxy for user *sip:bob@ethz.ch* is found, the INVITE message is forwarded (13). There is no difference in whether a SIP proxy receives an INVITE message from the local user or from another proxy over the network. Hence, *ProxyB*, upon receiving the INVITE message, checks whether the requested SIP URI is available in its local location service table (14). In this case it will find the entry (registered by Bob as 192.168.220.2:5062), and can then forward the INVITE message to Bob (15).

14.3 Internet-connected MANETs

In this section we enhance the architecture described in the previous Section to support Internet connectivity. Similar as with *DOPS*, we are interested in a seamless integration of *SIPHoc* with the SIP infrastructure of the Internet. Clients of SIPHoc should use their officially registered SIP accounts[2] transparently in the MANET. For instance, assume again a user Bob in the MANET with SIP URI *sip:bob@ethz.ch*. Given that Bob's SIP account is officially associated with the SIP provider at *ethz.ch*, we would like calls to and from the Internet to become possible as soon as the MANET has Internet access. On the other hand, Bob should always be able to call any SIP user within the MANET – and vice versa – even if the MANET is currently disconnected from the Internet.

We assume each *SIPHoc* instance to be in one of the following three states: *ad-hoc*,*Internet*, *hybrid*. In the *adhoc* state, where the only accessible network is the MANET, the *SIPHoc proxy* will operate as described in section 14.2. In the *Internet* state, where the only accessible network is the Internet, the proxy will act as a regular Internet-based SIP proxy. This means it will forward all the SIP messages to

---

[2] A SIP account associated with some official SIP provider in the Internet

their appropriate next hop SIP target. For instance a *SIP REGISTER* message of bob (*bob@ethz.ch*) would be forwarded to the SIP proxy that is responsible for the domain *ethz.ch*. If the *SIPHoc proxy* is in the *hybrid* state, where it has access to both the Internet and the MANET, the operation of the proxy is as follows. Besides registering users as described in section 14.2, the proxy now additionally forwards SIP REGISTER messages to the Internet if a responsible proxy for the specified domain is available. Upon receiving an INVITE message, the proxy contacts *MAND* to see if an outbound-proxy is available in the MANET for the given SIP target. If no outbound-proxy can be found, the INVITE message is forwarded to the Internet.

Typically, nodes with Internet access have multiple IP addresses assigned (see section 12.3.2). Let's call the IP address used in the MANET *internal* and the one used for communicating to the Internet *external*. SIP applications implement static binding, meaning that they use the IP address determined at startup time to be included in the contact address field of any SIP message sent towards the proxy (REGISTER, INVITE). Since the *external* IP address might be configured dynamically at the time the node gains Internet access, SIP messages would carry contact addresses pointing to an *internal* IP address. In practice, for connections to and from the Internet, one would however like to include the *external* IP address in the contact header of the SIP message because these addresses are used later by the application to establish the actual session. To take care of this issue, the *SIPHoc proxy*, upon receiving a SIP INVITE or REGISTER message, first saves the old contact address. It then changes the address in the SIP header of the message to match the external IP address. Once a response for a given request is received, the contact address is changed back to the original address that is used by the local SIP application. This keeps the application totally unaware of whether it communicates over the Internet or only within the MANET.

Figure 14.3 illustrates all the operations of the *SIPHoc* proxy described in this section. The activities, highlighted grey and labeled "Continue as usual", refer to the procedures described in section 14.2.

14.4 Case Study: VoIP in MANETs

Since SIPHoc is strictly SIP compatible, it allows out-of-the-box SIP based VoIP applications to run transparently in MANETs. In this section we use Kphone as a VoIP application to evaluate our SIP infrastructure. However, we also have successfully tested SIPHoc with various other Softphones such as Linphone, Twinkle or Ekiga. From a user perspective, the metric of interest is the *dial-to-ring* delay, i.e., the time elapsed between the caller clicking the button on the calling terminal and the time the called party hears the ringing. This call setup includes a set of SIP messages (INVITE, TRYING, RINGING) as well as the associated routing messages and a potential overhead of *MAND*. Figure 14.4 illustrates all the messages and time intervals that contribute to the dial-to-ring delay of a SIP call. The duration of the call setup depends on the current locations of caller and callee (MANET, Internet) and on the routing protocol used (AODV, OLSR). The following two sections evaluate the session setup time for a) pure Ad Hoc environments and b) Internet-connected Ad Hoc networks.

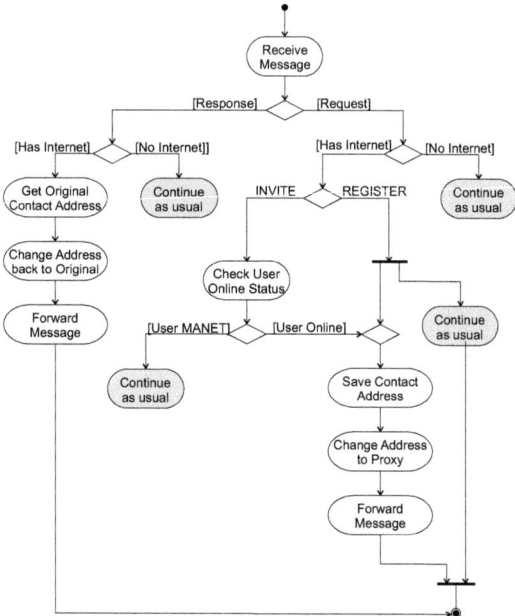

Figure 14.3: Activity Diagram of the *SIPHoc Proxy*

14.4.1 Experimental Setup

The measurements were done using 6 notebook computers running Debian 3.1 (Sarge). Five of them had a 2.0Ghz Mobile Pentium 4 and were equipped with an integrated 11Mbit/s IEEE802.11b wireless network interface card and were running kernel 2.6.8. The sixth laptop had a Pentium M processor with 1.73GHz combined with a 54Mbit/s IEEE802.11g wireless network interface card (used at 11Mbit/s through configuration). Similar as done when evaluating *DOPS* in chapter 13, we construct a multihop topology using packet filter rules. All results are average values computed over a set of 10 tests.

14.4.2 Performance in a MANET

In the first experiment we study the session setup time, once when using AODV and once when using OLSR as the MANET routing protocol. Figure 14.5a shows how the dial-to-ring delay evolves as the number of hopes between callee and caller grows. As a reference we also include the AODV route establishment time. Let's look at the dial-to-ring delay for AODV first. One can observe from Figure 14.5a that the SIP dial-to-ring delay is kept very low, only a few milliseconds more than the AODV route establishment time. This is because *MAND* provides an efficient mechanism to lookup the outbound-proxy target of a SIP INVITE message. The difference between the SIP dial-to-ring delay and the route establishment time comes

14.4. CASE STUDY: VOIP IN MANETS

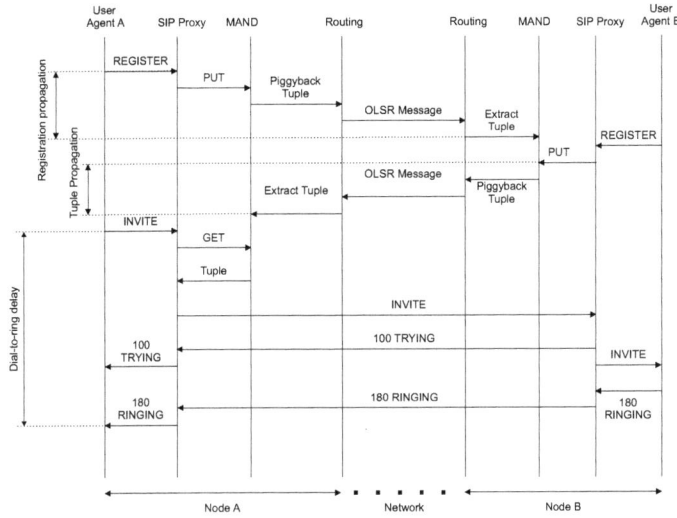

Figure 14.4: SIP registrations and call setup in a MANET using OLSR

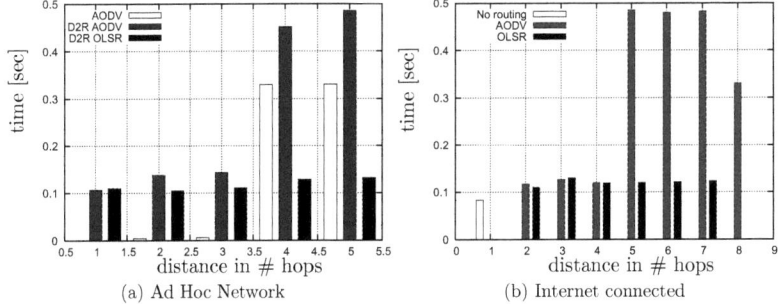

Figure 14.5: Dial-to-ring delay

from the three additional messages involved in a SIP session setup (INVITE, Trying, RINGING). Thus the measurements prove that SIPHoc reduces the dial-to-ring delay almost to the lowest value possible. The gap in time between three and four hops in Figure 14.5a is due to AODV's *expanded ring search* technique [93]. We have explained this phenomenon in section 11.5.5.

In contrast to the AODV case, the dial-to-ring delay for the OLSR case can be kept almost constant (around 0.1 seconds) and independent of the path length. This is because OLSR builds the routes in advance, before the SIP call takes place.

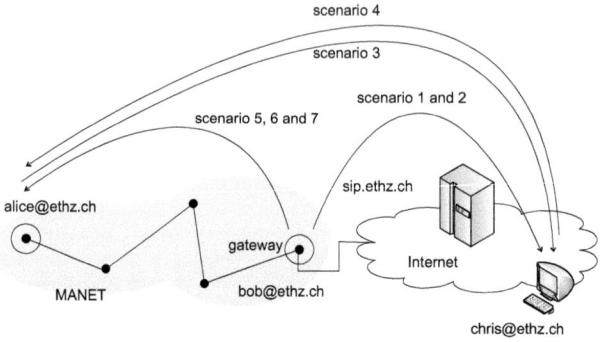

Figure 14.6: Evaluation setup

14.4.3 Internet-connected environment: scenarios

The next experiment evaluates the dial-to-ring delay for Internet connections. The setup consists of 5 laptops[3] arranged to form a linear 4-hop network. On the MANET side, there is user Alice with no direct Internet access. User Bob is on the other extreme of the MANET and is on a node that – in some experiments – acts as gateway and has Internet access. A third user, Chris, is located in the Internet. We also assume a SIP proxy on *sip.ethz.ch* located in the Internet. We have evaluated the *dial-to-ring* delay in 7 scenarios as illustrated in Figure 14.6: (1) Bob calling Chris using an unmodified out-of-the-box SIP proxy[4] as a forwarding proxy for Bob; (2) Bob calling Chris using SIPHoc; (3) Alice calling Chris using Bob as a gateway; (4) Chris calling Alice using Bob as a gateway; (5) Bob calling Alice when Bob is disconnected from the Internet; (6) Bob calling Alice when Bob is connected to the Internet; and (7) Bob calling Alice when Bob is connected to the Internet and Alice is connected to the Internet using the gateway provided by Bob (therefore the INVITE message is routed through the Internet). We also include the cost of an AODV route request (8) for comparison.

14.4.4 Internet-connected environment: experiments

Each set of bars in Figure 14.5b corresponds to the result of one scenario, with the x-axis representing the scenario identifier. For all scenarios except the first one, we consider both the case where AODV is used as a routing protocol and the case where OLSR is used.

Scenarios 1 and 2 are SIP connections over the Internet. They allow to determine the cost of SIPHoc over a plain SIP infrastructure. The results show that the overhead of SIPHoc (scenario 2) is small compared with the cost of using a plain SIP proxy (scenario 1). The overhead comes from the *SIPHoc Proxy* that performs various checks (looking up whether the node has Internet connection or not, contacting

---

[3]DELL Latitude, 2Ghz Intel Pentium IV, 256 MB RAM
[4]We have used the JAIN Proxy: http://snad.ncsl.nist.gov/proj/iptel

MAND) before forwarding the actual INVITE message to *sip.ethz.ch*. Thus, there is no difference between the dial-to-ring delay for AODV and OLSR in scenario 2.

Scenarios 3 and 4 are in the range of scenario 2 indicating that the 4-hop latency in the MANET is rather negligible compared to the whole session setup time. Furthermore, scenarios 3 and 4 confirm that both directions, MANET-Internet and Internet-MANET, perform similarly. The reason why there is no significant difference between the AODV and the OLSR case is that the route between Alice and the gateway has been established during the gateway discovery phase which took place asynchronously in a separate process (connection provider, Section 12.3.2) before the actual session establishment.

In scenario 5, the target user resides in the ad hoc network. If AODV is used, a route from Bob to Alice has to be established on demand which increases the dial-to-ring delay. One can observe that the overhead indeed is in the range of an AODV route establishment (scenario 8). If OLSR is used no route establishment overhead is observed due to the pro-active behavior of OLSR.

Scenario 6 leads to a similar result as scenario 5. Although Bob is connected to the Internet in scenario 6 the *SIPHoc* proxy is routing the SIP INVITE message directly towards the outbound-proxy it has discovered using *MAND*.

In scenario 7 both users Alice and Bob are connected to the Internet. However, the *SIPHoc* proxy always first contacts *MAND* for a potential outbound-proxy, thus routing the SIP INVITE message through the ad hoc network which results in a similar performance as scenario 5 and 6.

The results have shown that the overhead for calls to and from the Internet is within an acceptable range. This is mainly because outbound-proxy information for all users in the network are distributed pro-actively using *MAND*. Thus, deciding whether a given user is available in the ad hoc network or not comes at almost no cost.

14.5 Related Work

The Session Initiation Protocol SIP is a protocol set up a media session between two peers in the Internet. It is used most commonly in applications such as VoIP, video or chat. In a MANET, SIP must operate in a fully distributed manner while maintaining the same interface and procedures. This is a challenge because several of the problems caused by the decentralized nature of MANETS can be solved by ignoring SIP procedures. For instance, registrars can be eliminated if all applications pro-actively announce their contact addresses. This can be done using either a HELLO method [163] or REGISTER broadcast messages [164]. Such approaches create a significant message overhead and introduces incompatibilities with session mobility and tear-down. As an optimization, [164] suggests to use the Service Location Protocol (SLP) [133] to discover the SIP bindings. Unfortunately, SLP is also centralized and very inefficient in MANETs due to the heavy use of multicast [165]. Existing work to make SLP more efficient in MANETs is highly routing protocol specific [166].

SIP relies on DNS to locate nodes. A way to bypass the lack of a DNS infrastructure in MANETs is to restrict the network topology and assign specialized roles to given nodes [167]. This approach makes the endpoint discovery process easier and eliminates the need for (part of) the registrar since the position and communication paths to all nodes are known in advance. This approach does not introduce message overhead but imposes strong restrictions on the routing protocols. It is also very difficult to efficiently maintain a fixed routing topology in mobile settings.

Connecting the SIP-based applications to the Internet poses several problems. First, the SIP proxy does not know whether the target node is in the MANET or outside. Second, establishing sessions between nodes in and outside the MANET requires stable IP addresses. This can be done by using NAT [168] but it also requires additional mechanism like STUN [153]. If no NAT is available, extra mechanisms are needed to maintain consistent network addresses across connections of the MANET to the Internet. A way to avoid such problems is to assume the MANET is permanently connected to the Internet. Then one can impose a fixed network topology leading to the gateway [169]. This approach does not work for establishing SIP sessions in MANETs not connected to the Internet and re-introduces the problems of fixed topologies.

14.6 Summary

In this chapter we have presented *SIPHoc*, a middleware providing SIP-compatible session establishment in both isolated and Internet-connected MANETs. Our experiments illustrate that SIPHoc is able to establish sessions within a few hundred milliseconds, regardless whether the other party is located in the MANET or in the Internet. Due to the efficient architecture of SIPHoc, the overhead in the dial-to-ring delay stays in the order of a route discovery time, which is the lowest value possible since the cost of a route discovery has to be paid anyway.

Chapter 15

Network services in MANETs: A Discussion

The network services described in the previous chapters share some important design principles and system properties. We want to briefly highlight those and reflect about their impact onto different quality metrics of the system.

1. **Proxy architecture:** In a MANET, network services must operate in a distributed manner while maintaining compatibility with the standard they are implementing. We adopted a proxy-based approach to meet both requirements. The basic concept is to run a proxy of a given service on each node in the MANET and share common objects (such as, e.g., contact information of SIP users) using *MAND*. The local proxy makes it possible to transparently hide the distributed implementation from the standard interface provided to upper layers.

2. **Piggybacking:** Turning a centralized or hierarchical network service into a distributed service for MANETs requires distributing service data across the network. MANETs are known to have only limited throughput capacity. Through the use of routing message piggybacking we were able to come up with a solution that is highly message efficient.

3. **Pro-active:** Applications like chat or VoIP may send many DNS or SIP requests during a regular session. The response time of those requests directly determines how fast those applications can react to user events, such as, session establishment or message exchange. Thus, a fast response time must be an important criteria for new service architectures considering that communication delays in MANETs are already high compared to, e.g., the Internet. All our services implement a *pro-active* distribution of service data. By doing so, a given service is able to respond immediately to a request by consulting his local cache. This is of particular importance when interacting with the Internet. If local information can used to determine whether a given service can be provided in the MANET or not, a network service will be able to immediately forward a request to the Internet without noticeable overhead.

4. **Lazy distribution:** Typically, pro-active approaches have been shown to be message inefficient. This is true for instance in terms of routing. We already

discussed the fact that our services use message piggybacking to avoid additional traffic load. However, what saves all those extra messages is not only the piggybacking, but also the fact that service data is sent *lazy*. That way we are able to multiplex various service data of different services onto one single packet. *Lazy* data distribution has been shown to be very efficient also in other fields of research. For instance, it has been shown that a *delay-tolerant* mechanism to exchange packets in wireless networks can lead to a significant improvement in throughput capacity [170]. In the field of database replication, *lazy* data synchronization has successfully been used to decrease the transaction response time [171].

5. **Reliability:** Network services such as DNS or SIP have to operate as reliable as possible. The limiting factor thereby is the network itself. In the Internet, the network is supposed to be reliable, thus service access is possible all the time. In a MANET, the network is inherently unreliable. The wireless signal is error prone and packet loss is a common observation. Those problems induce further difficulties on higher layers such as routing. Routing is an essential component of the IP layer which all the network services described in this thesis are based on. Failing to establish a route between to IP endpoints makes it impossible to access a given service remotely. Thus, routing constitutes the limiting factor for all network services in MANETs. By piggybacking service data onto routing messages, we interlink the service data availability with the ability to access a service. Or in other words, if the service data is not available to the service, then this is because there is currently no route between the producer and the requester of the service data.

6. **Memory consumption:** Some of the network services we presented implement full replication of service data. Therefore, one issue to be considered is memory consumption. In *DOPS* the information that is stored on each node is the domain names and the IP addresses of each of the other nodes in the network. Since each node typically has only one wireless interface with a configured hostname assigned, the amount of data to be stored at each node is a linear function of the total number of nodes in the network. If we assume a directory mapping (domain name, IP address) can be stored using, on average, 50 bytes (4 bytes for the IP address and 46 bytes for the domain name), 25K of memory would be sufficient to support a network of 500 nodes. In the case of *SIPHoc*, we can assume a similar upper bound for the memory consumption since it is unrealistic to have more than one active SIP account per node. For *MANET-SLP* the amount of memory needed grows linearly with the number of services that are registered. There might be more than one service per node. The only way to handle this issue is to define an upper limit on the total number of services that can be registered on one single node.

Most systems dealing with storing and looking up data have to deal with tradeoffs among several properties, like response time, update time, consistency, memory space, etc. In most cases it is impossible to build win-win systems, for instance providing fast response and update times, offering high consistency, and consuming less memory. When building a system, one has to optimize for a subset of these properties. Which properties to optimize for depends on the characteristics of the system (e.g., the ratio between queries and updates) and the context in which the system is supposed to run. The decision made in this thesis, when designing the various network services, were influenced by the intrinsic properties of MANETs. Some of those properties, like connectivity or throughput capacity, have been studied in the first part of this thesis.

Chapter 16
Social Ad Hoc Networking

As a fourth an final demonstrator of *MAND* we present in this chapter *AdSocial*, a social networking application for MANETs. *AdSocial* allows nearby users in the MANET to interact spontaneously by exchanging profile information or by opening a chat or VoIP conversation. We have deployed *AdSocial* on a testbed of 23 handheld devices and tested it during various group activities.

16.1 Overview

People use social networks to exchange information, images, video clips or to just communicate common interests. In contrast to many sites providing this kind of interaction over the Internet, like Facebook, ICQ, etc., *AdSocial* runs on MANETs. The key difference between *AdSocial* and traditional social networks lies in the way users communicate with each other. In Internet-based social applications users communicate with so called friends (other previously contacted users). In *AdSocial*, interaction takes place between nearby users in the ad hoc network, so called *buddies*. An ad hoc network might emerge temporarily in everyday's life or during business, e.g., on the train when travelling to work, at a conference or during a business meeting. An *AdSocial* user can create a profile in which he/she specifies interests and other personal information. *AdSocial* signals the user's presence in the ad hoc network using *MAND*. The users in the ad hoc network can retrieve the profile of their buddies as well as search for nearby buddies with specific interests. Moreover, *AdSocial* provides the basis for running collaborative applications. For instance, *AdSocial* integrates with existing VoIP or chat applications on the device. If a profile of a user contains a SIP VoIP address or a chat address, *AdSocial* will immediately start a VoIP or chat session with the given person. In all those cases, *AdSocial* is used as a first step to get in contact with people around each other.

16.2 Architecture

AdSocial runs as a local web application in a regular browser. The various components involved on one single node in the ad hoc network are illustrated in Figure 16.1. The local *browser* connects to the *web server* that redirects the call to the *AdSocial* application via a *FastCGI* interface [172]. Depending on the state of the web application *AdSocial* serves static files containing the presentation logic or gathered

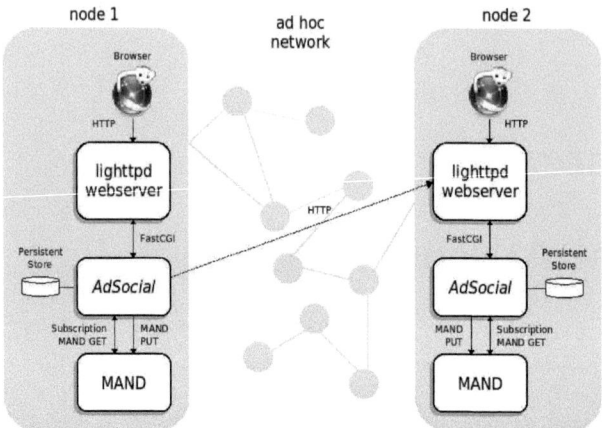

Figure 16.1: The *AdSocial* system architecture

information about nearby buddies. The profile of the user containing his or her name, contact information, etc. is stored persistently in a file.

16.2.1 Presentation logic

The whole presentation logic is written in JavaScript and executed by the browser. The interaction between the presentation logic and *AdSocial* is done using *AJAX* polling. In a regular interval, the browser queries the local *AdSocial* instance to get updates about the availability of nearby users.

One characteristic of *AJAX* is its disability to interact with a webserver that is different from the one that has served the web page in the first place. This is due a restriction inflicted by the browser. As a workaround, the *AdSocial* FastCGI module offers bridging functionality to redirect remote *AJAX* requests. The bridging functionality is used by *AdSocial* to retrieve the profile of a nearby user.

16.2.2 Buddy discovery

AdSocial advertises the presence of the currently logged in user with *MAND* by periodically issuing a *MAND insert* call. The tuple inserted into *MAND* for a user Bob may look as follows:

```
tuple{
    key=adsocial,
    value=1.2.3.4:80/gsn/xml.gsn&Bob1,
    scope=64,
    version=4,
    lifetime=0
}
```

The *value* of the tuple contains information about the IP address of local node, the port where the webserver is running, the absolute path to the XML engine on local web server, the actual nickname of the user and its availability status. The availability status is specified by the user and corresponds to either *available* (1), *busy* (2) or *away* (2). Tuples describing user presence information are issued with a *lifetime* of zero and a large *scope*. Tuples of this form can be seen as events notifying remote applications with open subscriptions for the common event/tuple key ("adsocial"). Other examples of *MAND* events are given in chapter 12 (SLP) and in Figure 11.3.

Users in the context of a list of friends are often referred to as "buddies". The list of buddies of a user is maintained by *AdSocial* based on the *MAND* events it receives. In order for *AdSocial* to receive all events containing buddy presence information it will periodically issue a *MAND lookup* call (see section 11.4 for details).

```
request{
    key=adsocial,
    scope=0,
    version=0,
    lifetime=60
}
```

16.2.3 VoIP and video: integration with other applications

One of the most exciting features of *AdSocial* is that it integrates very well with other collaborative applications. Since the user interface runs in a browser, the browser can be configured to perform an associated action for a specific file type. We have leveraged this feature to interlink *AdSocial* with VoIP and video applications using our *SIPHoc* infrastructure. Each device running *AdSocial* will additionally have *SIPHoc* and a VoIP client installed. If a user retrieves a profile of another nearby buddy, and the profile contains the SIP address of the buddy, he/she will be able to open a VoIP session by just clicking on the SIP address within the profile.

Similarly as for VoIP and video, one could also configure the browser to interlink *AdSocial* with an instant messenger. However, *AdSoical* already integrates a simple chat application to support chat among buddies in the ad hoc network.

16.3 Deployment

AdSocial was designed and implemented for Linux, as all other systems that have been presented in this thesis. Since *AdSocial* is supposed to be a pervasive application it is important to deploy it on devices people can carry with them in their everyday life. We have deployed *AdSocial* on *Nokia Internet Tablets* [173]. Those devices run Linux (Maemo Linux [174]), are small enough to be carried around and they are equipped with a WLAN card that can be switched into ad hoc mode. The upper part of Figure 16.2 shows two Nokia N800 Internet tables with *AdSocial* running. The figures show the user profiles of two users, Patrick and Oriana. The left hand side of the profiles contains personal information of the users, such as a photo, his or here interests or the SIP address. On the right hand side the nearby buddies are displayed. Users can retrieve the profile of nearby buddies by right-clicking on the buddy's icon. Similarly, they can also open a chat session with a buddy.

146 CHAPTER 16. SOCIAL AD HOC NETWORKING

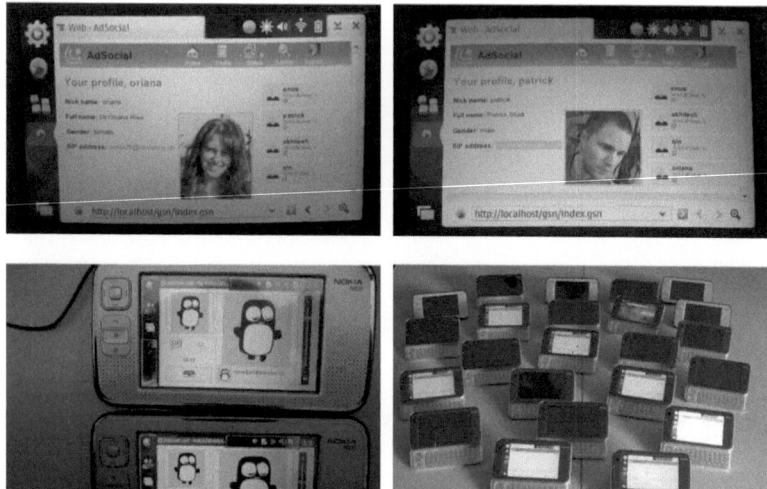

Figure 16.2: AdSocial: social networking among Nokia N810 devices

Each Nokia device by default comes with a VoIP and video application. Moreover, the standard web browser on the devices is already configured to use the VoIP/video application when opening SIP URL's. Hence, *AdSocial* users on the Nokia devices that have their SIP address specified in their profile can easily initiate VoIP or video conversations among themselves by clicking on other user's SIP address. A picture of a phone conversation between two users in the ad hoc network is shown in the lower left corner of Figure 16.2. What cannot be seen from the picture is that the VoIP application displays "Internet call" whenever a call is coming in from another user in the ad hoc network. This once more demonstrates that *SIPHoc* is SIP compatible letting SIP applications believe that they are communication across the Internet, while they are actually communication over an ad hoc network.

Experiments with social applications require a minimum amount of users to participate. Nokia kindly provided us with 20 N810 devices to be used for larger experiments. A picture of all devices is shown in the right corner of Figure 16.2.

16.4 Measurements

AdSocial was tested during the *Retreat* 2008 of the ETH Systems group, held in Arosa, a small town in the mountains of Switzerland. The event brought together 34 members of the group for three days in total. Our testbed consists of 18 Nokia N810 Internet tables, which have been distributed among the attendants of the event. We have used OLSR as the MANET routing protocol since earlier tests have shown that OLSR performs better than AODV in terms of route stability and route recovery.

16.4. MEASUREMENTS

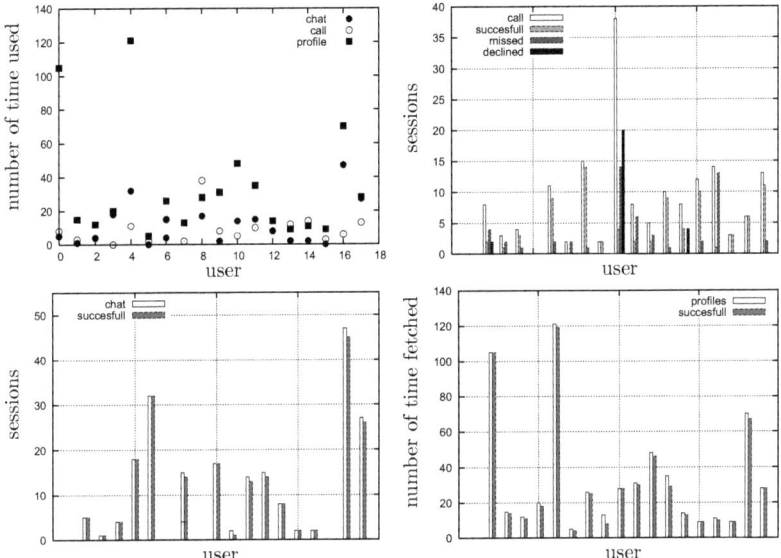

Figure 16.3: User behavior of a case study during several group activities

16.4.1 User statistics

In this section we present some statistics about the use of *AdSocial* in general. We study, for each component of the application – browsing other people's profile, chat, VoIP – how many times each user has used the particular component and how many of those attempts were successful. In Figure 16.3a we see an overall comparison of which components have been used more frequently by the user. One can observe that people did use all of the components in a good mixture, however, browsing through other people's profile clearly has been the most popular activity. This makes sense, since the profile of a *buddy* not only contains personal interests but also some address information, e.g. the SIP address, that can be used to establish a chat or a phone conversation.

Figure 16.3b shows some more detailed information regarding SIP VoIP calls that have been made during the three days of evaluation. The figure shows on a per user basis the number of SIP calls initiated, the number of successful calls, the number of missed calls, and the number of declined calls. We can see that most calls were either successful or missed. The reason why some calls were missed might be that people were new to this form of communication and also had to learn the handling of the devices. Similar as done for the SIP calls, Figure 16.3c compares for each user the number of chat attempts to the number of chat sessions that were successful. Figure 16.3d illustrates the same statistics but for profile retrievals. From the figures one can observe that the number of failed chat sessions or failed profile retrievals are considerably low with respect to the total amount of requests.

CHAPTER 16. SOCIAL AD HOC NETWORKING

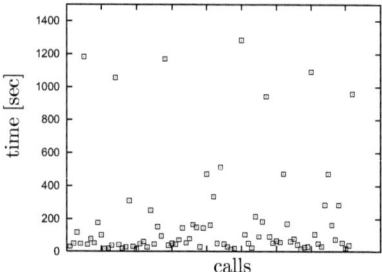

Figure 16.4: SIP call durations

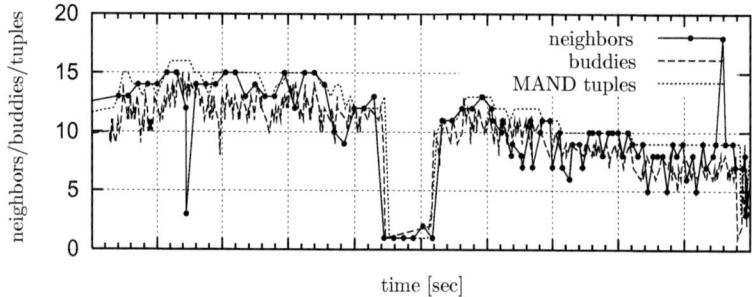

Figure 16.5: SIP call durations and buddies/neighbors/tuples vs time

16.4.2 VoIP statistics

As illustrated in Figure 16.2a, during the three days of evaluation people made extensive use of the VoIP and video applications which are interlinked with *AdSocial* through the browser (Section 16.2.3). Figure 16.4 shows the durations of each call made. Every point in Figure 16.4 corresponds to a phone conversation with its y-value describing the duration of the call measured in seconds. As can be observed, most phone calls lasted less than 100 seconds, however, there are also some calls with a duration of more than 16 minutes. It's very likely that the long calls were actually video conversations where the devices were used like a web-cam and were not moved for most of the time.

In Figure 16.5 we study different properties related to *AdSocial*, *SIPHoc* and *MAND*. The Figure shows the overall correlation between all the SIP tuples, number of *AdSocial* buddies, and number of neighbors in the routing table. The graph has been derived from the log files of one device over a period between 12:45 and 15:30, which is a period when most devices were in frequent use. The graph shows that the variation of the number of neighbors is much higher than the variation of the number of buddies, which is still higher than the variation of the number of SIP tuples. The fact that the reactivity of *MAND* does not entirely map the constant fluctuations in the routing table is due to the lifetimes of the tuples. Remember that tuples have a specified *lifetime* for which they are guaranteed to stay in the system, even if they

16.4. MEASUREMENTS

are not refreshed. Thus, a tuple may still be in the system although the node that originally generated the tuple has already left the network. From the Figure we also observe that if the number of buddies change over a longer period of time, the tuples in the *MAND* system will change accordingly.

There is no project that addresses social networks in MANETs. But there are some social network related approaches that are worth mentioning. The Internet company Google is about to define a standard for storing, retrieving and processing user related data in social networks. The project is called *OpenSocial* [175] and is still in alpha state. It mainly describes a JavaScript API for manipulating profile data. With the interface it will be possible to create new services using the existing infrastructure of an social network provider. The problem up to now is that OpenSocial is only a collection of functions that a social network provider should implements to allow the community to grow with the ideas of its members. Letting the member enhance the functionality of the service. There is no reference implementation and the definitions of what the functions should do are sufficiently general that no final implementation has been derived so far.

The OpenID Foundation provides a mechanism to maintain a user profile in a central place and powering other sites using such data with the deposited information. The project is called *OpenID*[176]. Whenever a user joins a message board or a community site he or she has to state certain data, mainly contact information like name and e-mail address. But this has to be done for each such site over and over again. OpenID collects this data once. Then the user can join the OpenID enabled community and the required data is transmitted to the site. The user only needs to allow the community server to access his or her profile. OpenID allows to maintain the profile in one central place, but also updates and changes to the profile can be delegated to the connected community sites depending on their implementation. OpenID also permits the social network applications to collect data locally. They are not forced to request the data again. This is why the update mechanism may fail. But sites that use OpenID for more than just a login replacement should take care of that.

During the last years most research addressing "ad hoc" and "social networks" has been targeting the development of appropriate models to simulate mobility [177] in MANETs. Heavy load and behaviour is tested not in real life but with simulations. Thus analysis that model that and other interesting behaviour [178] more accurately were and are developed. In the past many such models evolved because of findings based on *small world* [179] observations of MANETs. Thus the models may have been too optimistic, because the small world phenomena does not apply to all wireless networks. But it may still be realistic, because social networks tend to be not as general as ad hoc networks could be [180].

There is a trend to port existing social network ideas to mobile phone architectures. Depending on the approach they use the web to access contents or run entirely message based. *Mobikade* [181] is a Web based free mobile social networking service (SNS) aiming to bring the massive success of Japanese mobile SNS sites to Europe. It differentiates from other social networking sites by the fact that it is an exclusively mobile based community and accessible only through the mobile phone. *Dodgeball* [182] is an SMS based social networking software provider for mobile devices. Users text their location to the service, which then notifies them of crushes, friends, friends' friends and interesting venues nearby. Many others like Zingku, Rabble, Moblabber, Wadja, Zemble, twitter, etc. work all similarly. More or less they are just group messaging systems for mobile phones with some additional features.

There are some implementations of social network applications that work on mobile phones and other portable devices that are not message or web based, but they mainly use bluetooth to communicate. Some examples are MobiLuck, ProxiDating, CrowdSurfer. There is also a relatively new application that uses a wireless LAN connection between the devices to detect nearby users. It is called Jambo Networks. Jambo is also a group communication tool that will detect members of the same group or groups the current user is participating. It is also possible to search for people in the area and contact them.

16.5 Summary

In this chapter we presented the design and implementation of a social networking application that addresses the issues of MANETs. Users are able to create profiles describing their interests and contact information, search for other users with similar interests and even to directly contact each other with the small built in chat application. Moreover, *AdSocial* integrates MANET VoIP and allows users to seamlessly set up a phone call among each other. We have deployed *AdSocial* on Nokia N800 handheld devices and evaluated the application during a group activity in the mountains of Switzerland.

Chapter 17
Conclusion

Mobile Ad Hoc Networks have not yet become reality. Most of the deployments are either of very small scale or they are part of academic research projects. The problems to be solved for MANETs to enter real life are of various forms. In this dissertation we have addressed two sets of problems which we consider to be particularly important. In the first part of the thesis we have studied fundamental properties of ad hoc networks. We followed a numerical approach based on Monte-Carlo methods which allowed us to put a strong emphasize on realistic physical layer models. As one example, we showed how network connectivity and throughput capacity evolves under log-normal shadowing radio propagation. Our results illustrate that the radio irregularity induced by log-normal shadowing has indeed a beneficial impact on connectivity and throughput capacity. This is of interest since it shows that an analysis of connectivity and interference in a circular radio propagation model yields worst case bounds for both connectivity and interference. Until now, the assumption that shadowing increases interference led to assuming that a circular radio propagation range was a best case scenario for network capacity. As part of our effort to consider realistic network models we have further studied the impact of different interference models on throughput capacity. It was shown that there is a significant gap between throughput capacity under the protocol model versus under the physical model. This is of importance since the protocol model served as the basis for many analytical results on capacity and interference. In the last chapter of part I we have extended our approach on throughput capacity to study bandwidth reservation in ad hoc networks. More precisely, we show that there is an inherent trade-off between the quality of a bandwidth reservation and the amount of resources wasted due to over-reservations. The quality of the reservation deteriorates even more when considering irregular radio propagation. Overall, the first part of this thesis presents interesting new insight into the behavior of fundamental properties of ad hoc networks under realistic physical layer models and finite network boundaries. Taken together, the results may help to assess and design real network deployments.

In the second part, we considered ad hoc networks from a system's research perspective. We have presented several network abstractions and services that facilitate the seamless transfer of standard IP-based applications to MANETs. The virtual interface described in chapter 10 allows nodes with different media access technologies to form one single IP based mobile ad hoc network. In such a network, routing, packet forwarding and mobility support take place independent of whatever wireless access technology is used on a device. We belief that our approach of virtualizing the communication layer not only provides an efficient solution to integrate heterogeneous classes of devices and network stacks, but is also an important technique to build any form of distributed service in MANETs. We have presented in chapters 12, 13 and 14 distributed SLP, DNS and SIP services, each of them appearing as one virtual network service to the application layer. Those services allow standard Internet-based applications like, e.g., VoIP or chat, to run in MANETs without need of any modification. Moreover, the network services integrate seamlessly with the Internet and therefore provide applications with the possibility to communicate to and from the Internet whenever one node in the MANET has Internet access. The core part of the services presented in this thesis relies on *MAND* (chapter 11), an efficient distributed key/value store. *MAND* is message efficient since the distribution of key/value information is solely based by piggybacking onto routing messages, traffic that is sent anyway. As a fourth and final demonstrator of *MAND*, we presented a novel social networking application for MANETs. *AdSocial* allows to browse personal profiles of nearby users and provides abilities open a chat or VoIP session in case there is interest. All of the systems presented in part II of the thesis have been deployed on Nokia N800 handheld devices and approved by real experiments. Overall, the second part of this thesis demonstrates how a middleware based approach can be used to build a transparent platform for existing networking applications to be run in MANETs. In this regard, our approach can be seen as a new paradigm for distributed systems in MANETs, in contrast to former approaches where dedicated applications have been considered to be the solution to cope with the complexity of MANETs.

Bibliography

[1] Y.-C. Cheng and T.G. Robertazzi. Critical connectivity phenomena in multi-hop radio models. *Communications, IEEE Transactions on*, 37(7):770–777, Jul 1989. ISSN 0090-6778. doi: 10.1109/26.31170.

[2] P. Gupta and P. Kumar. Critical power for asymptotic connectivity in wireless networks. *Stochastic Analysis, Control, Optimization and Applications*, 1998.

[3] O. Dousse and P. Thiran. Connectivity vs capacity in dense ad hoc networks. In *O. Dousse and P. Thiran, Connectivity vs capacity in dense ad hoc networks, in Proc. of IEEE INFOCOM, Hong Kong, 2004.*, 2004. URL citeseer.ist.psu.edu/dousse04connectivity.html.

[4] O. Dousse, P. Thiran, and M. Hasler. Connectivity in ad-hoc and hybrid networks. *INFOCOM 2002. Twenty-First Annual Joint Conference of the IEEE Computer and Communications Societies. Proceedings. IEEE*, 2:1079–1088 vol.2, 2002. ISSN 0743-166X. doi: 10.1109/INFCOM.2002.1019356.

[5] Paolo Santi, Douglas M. Blough, and Feodor Vainstein. A probabilistic analysis for the range assignment problem in ad hoc networks. In *MobiHoc '01: Proceedings of the 2nd ACM international symposium on Mobile ad hoc networking & computing*, pages 212–220, New York, NY, USA, 2001. ACM Press. ISBN 1-58113-428-2.

[6] Christian Bettstetter. On the minimum node degree and connectivity of a wireless multihop network. In *MobiHoc '02: Proceedings of the 3rd ACM international symposium on Mobile ad hoc networking & computing*, pages 80–91, New York, NY, USA, 2002. ACM Press. ISBN 1-58113-501-7. doi: http://doi.acm.org/10.1145/513800.513811.

[7] Madhav Desai and D. Majunath. On the connectivity in finite ad hoc networks. *IEEE Communications Letters*, 2002.

[8] Thomas Philips, Shivendra Panwar, and Asser Tantwani. Connectivity properties of a packet radio network model. *IEEE Transactions on Information Theory*, 1989.

[9] Philippe Piret. On the connectivity of radio networks. *IEEE Transactions on Information Theory*, 1991.

[10] Ao Tang, Cedric Florens, and Steven Low. An empirical study on connectivity of ad hoc networks. *Proceedings of IEEE Aerospace Conference*, 2003.

[11] Marco Zuniga and Bhaskar Krishnamachari. Analyzing the transitional region in low power wireless links. *First International Conference on Sensor and Ad Hoc Communications and Networks (SECON)*, 2004.

[12] Gang Zhou, Tian He, Sudha Krishnamurthy, and John A. Stankovic. Impact of radio irregularity on wireless sensor networks. In *MobiSys '04: Proceedings of the 2nd international conference on Mobile systems, applications, and services*, pages 125–138, New York, NY, USA, 2004. ACM Press. ISBN 1-58113-793-1. doi: http://doi.acm.org/10.1145/990064.990081.

[13] Michele Zorzi and Silvano Pupolin. Optimum transmission ranges in multihop packet radio networks in the presence of fading. *IEEE Trans. On Communication*, 43(7), July 1995.

[14] Olivier Dousse, François Baccelli, and Patrick Thiran. Impact of interferences on connectivity in ad hoc networks. *IEEE/ACM Trans. Netw.*, 13(2):425–436, 2005. ISSN 1063-6692. doi: http://dx.doi.org/10.1109/TNET.2005.845546.

[15] Daniel Aguayo, John Bicket, Sanjit Biswas, Glenn Judd, and Robert Morris. Link-level measurements from an 802.11b mesh network. *SIGCOMM Comput. Commun. Rev.*, 34(4):121–132, 2004. ISSN 0146-4833. doi: http://doi.acm.org/10.1145/1030194.1015482.

[16] Christian Bettstetter and Christian Hartmann. Connectivity of wireless multihop networks in a shadow fading environment. In *MSWIM '03: Proceedings of the 6th ACM international workshop on Modeling analysis and simulation of wireless and mobile systems*, pages 28–32, New York, NY, USA, 2003. ACM Press. ISBN 1-58113-766-4. doi: http://doi.acm.org/10.1145/940991.940998.

[17] P. Gupta and P. Kumar. Capacity of wireless networks. *Information Theory*, 46(2):388–404, 2000.

[18] Pradeep Kyasanur and Nitin H. Vaidya. Capacity of multi-channel wireless networks: impact of number of channels and interfaces. In *MobiCom '05*, New York, NY, USA, 2005. ACM Press. ISBN 1-59593-020-5. doi: http://doi.acm.org/10.1145/1080829.1080835.

[19] Jinyang Li, Charles Blake, Douglas S.J. De Couto, Hu Imm Lee, and Robert Morris. Capacity of ad hoc wireless networks. In *MobiCom '01*, New York, NY, USA, 2001. ACM Press. ISBN 1-58113-422-3. doi: http://doi.acm.org/10.1145/381677.381684.

[20] Andrew Brzezinski, Gil Zussman, and Eytan Modiano. Enabling distributed throughput maximization in wireless mesh networks: a partitioning approach. In *MobiCom '06: Proceedings of the 12th annual international conference on Mobile computing and networking*, pages 26–37, New York, NY, USA, 2006. ACM Press. ISBN 1-59593-286-0. doi: http://doi.acm.org/10.1145/1161089.1161094.

[21] Alireza Keshavarz-Haddad, Vinay Ribeiro, and Rudolf Riedi. Broadcast capacity in multihop wireless networks. In *MobiCom '06: Proceedings of the*

12th annual international conference on Mobile computing and networking, pages 239–250, New York, NY, USA, 2006. ACM. ISBN 1-59593-286-0. doi: http://doi.acm.org/10.1145/1161089.1161117.

[22] Gaurav Sharma, Ravi R. Mazumdar, and Ness B. Shroff. On the complexity of scheduling in wireless networks. In *MobiCom '06: Proceedings of the 12th annual international conference on Mobile computing and networking*, pages 227–238, New York, NY, USA, 2006. ACM Press. ISBN 1-59593-286-0. doi: http://doi.acm.org/10.1145/1161089.1161116.

[23] Hyuk Lim, Chaegwon Lim, and Jennifer C. Hou. A coordinate-based approach for exploiting temporal-spatial diversity in wireless mesh networks. In *MobiCom '06: Proceedings of the 12th annual international conference on Mobile computing and networking*, pages 14–25, New York, NY, USA, 2006. ACM Press. ISBN 1-59593-286-0. doi: http://doi.acm.org/10.1145/1161089.1161093.

[24] Thomas Moscibroda, Roger Wattenhofer, and Aaron Zollinger. Topology control meets sinr:: the scheduling complexity of arbitrary topologies. In *MobiHoc '06: Proceedings of the seventh ACM international symposium on Mobile ad hoc networking and computing*, pages 310–321, New York, NY, USA, 2006. ACM Press. ISBN 1-59593-368-9. doi: http://doi.acm.org/10.1145/1132905.1132939.

[25] Pablo Soldati, Björn Johansson, and Mikael Johansson. Proportionally fair allocation of end-to-end bandwidth in stdma wireless networks. In *MobiHoc '06: Proceedings of the seventh ACM international symposium on Mobile ad hoc networking and computing*, pages 286–297, New York, NY, USA, 2006. ACM Press. ISBN 1-59593-368-9. doi: http://doi.acm.org/10.1145/1132905.1132937.

[26] Olga Goussevskaia, Yvonne Anne Oswald, and Roger Wattenhofer. Complexity in geometric sinr. In *MobiHoc '07: Proceedings of the 8th ACM international symposium on Mobile ad hoc networking and computing*, pages 100–109, New York, NY, USA, 2007. ACM. ISBN 978-1-59593-684-4. doi: http://doi.acm.org/10.1145/1288107.1288122.

[27] Gurashish Brar, Douglas M. Blough, and Paolo Santi. Computationally efficient scheduling with the physical interference model for throughput improvement in wireless mesh networks. In *MobiCom '06: Proceedings of the 12th annual international conference on Mobile computing and networking*, pages 2–13, New York, NY, USA, 2006. ACM Press. ISBN 1-59593-286-0. doi: http://doi.acm.org/10.1145/1161089.1161092.

[28] D. Chafekar, V.S.A. Kumart, M.V. Marathe, S. Parthasarathy, and A. Srinivasan. Approximation algorithms for computing capacity of wireless networks with sinr constraints. *INFOCOM 2008. The 27th Conference on Computer Communications. IEEE*, pages 1166–1174, April 2008. ISSN 0743-166X. doi: 10.1109/INFOCOM.2008.172.

[29] Olga Goussevskaia, Roger Wattenhofer, Magnus Halldorsson, and Emo Welzl. Scheduling arbitrary wireless networks. *INFOCOM 2009. Twenty-eight AnnualJoint Conference of the IEEE Computer and Communications Societies*, 2009.

[30] Theodore S. Rappaport. *Wireless Communications: Principles & Practice*. Prentice Hall, 2002.

[31] Weizhao Wang, Xiang-Yang Li, Ophir Frieder, Yu Wang, and Wen-Zhan Song. Efficient interference-aware tdma link scheduling for static wireless networks. In *MobiCom '06: Proceedings of the 12th annual international conference on Mobile computing and networking*, pages 262–273, New York, NY, USA, 2006. ACM Press. ISBN 1-59593-286-0. doi: http://doi.acm.org/10.1145/1161089.1161119.

[32] J. Gomez and A.T. Campbell. A case for variable-range transmission power control in wireless multihop networks. *INFOCOM 2004. Twenty-third AnnualJoint Conference of the IEEE Computer and Communications Societies*, 2: 1425–1436 vol.2, March 2004. ISSN 0743-166X.

[33] Tae-Suk Kim, Jennifer C. Hou, and Hyuk Lim. Improving spatial reuse through tuning transmit power, carrier sense threshold, and data rate in multihop wireless networks. In *MobiCom '06: Proceedings of the 12th annual international conference on Mobile computing and networking*, pages 366–377, New York, NY, USA, 2006. ACM Press. ISBN 1-59593-286-0. doi: http://doi.acm.org/10.1145/1161089.1161131.

[34] Ahmed Bader and Eylem Ekici. Throughput and delay optimization in interference-limited multihop networks. In *MobiHoc '06: Proceedings of the seventh ACM international symposium on Mobile ad hoc networking and computing*, pages 274–285, New York, NY, USA, 2006. ACM Press. ISBN 1-59593-368-9. doi: http://doi.acm.org/10.1145/1132905.1132936.

[35] Feng Xue, Liang-Liang Xie, and Panganamala R. Kumar. The transport capacity of wireless networks over fading channels. *IEEE Transactions on Information Theory*, 51(3):834–847, 2005. URL http://dx.doi.org/10.1109/TIT.2004.842628.

[36] Peter Barbrand and Di Yuan. Maximal throughput of spatial tdma in ad hoc networks. In *3rd Scandinavian Workshop on Wireless Ad-hoc Networks*, 2003.

[37] P. Bjorklund, P. Varbrand, and Di Yuan. Resource optimization of spatial tdma in ad hoc radio networks: a column generation approach. *INFOCOM 2003. Twenty-Second Annual Joint Conference of the IEEE Computer and Communications Societies. IEEE*, 2:818–824 vol.2, March-3 April 2003. ISSN 0743-166X.

[38] Stavros Toumpis and Andrea Goldsmith. Capacity regions for wireless adhoc networks, 2001. URL citeseer.ist.psu.edu/toumpis01capacity.html.

[39] R. Braden, D. Clark, and S. Shenker. Integrated Services in the Internet Architecture: an Overview. RFC 1633 (Informational), June 1994. URL http://www.ietf.org/rfc/rfc1633.txt.

[40] S. Blake, D. Black, M. Carlson, E. Davies, Z. Wang, and W. Weiss. An Architecture for Differentiated Service. RFC 2475 (Informational), December 1998. URL http://www.ietf.org/rfc/rfc2475.txt. Updated by RFC 3260.

BIBLIOGRAPHY

[41] G. Ahn, A. Campbell, A. Veres, and L. Sun. Swan: Service differentiation in stateless wireless ad hoc networks. *IEEE INFOCOM'2002.*, 2002.

[42] R. Braden, L. Zhang, S. Berson, S. Herzog, and S. Jamin. Resource ReSerVation Protocol (RSVP) – Version 1 Functional Specification. RFC 2205 (Proposed Standard), September 1997. URL http://www.ietf.org/rfc/rfc2205.txt. Updated by RFCs 2750, 3936.

[43] Anup Kumar Talukdar, B. R. Badrinath, and Arup Acharya. Mrsvp: a resource reservation protocol for an integrated services network with mobile hosts. *Wirel. Netw.*, 7(1):5–19, 2001. ISSN 1022-0038. doi: http://dx.doi.org/10.1023/A:1009035929952.

[44] Chien-Chao Tseng, Gwo-Chuan Lee, Ren-Shiou Liu, and Tsan-Pin Wang. Hmrsvp: a hierarchical mobile rsvp protocol. *Wirel. Netw.*, 9(2):95–102, 2003. ISSN 1022-0038. doi: http://dx.doi.org/10.1023/A:1021833430898.

[45] J. Manner and K. Raatikainen. Extended quality-of-service for mobile networks. *IWQoS*, 2001.

[46] Bongkyo Moon and H. Aghvami. Reliable rsvp path reservation for multimedia communications under an ip micromobility scenario. *IEEE Wireless Communications*, 9(5):93–99, Oct 2002.

[47] M. Mirhakkak, N. Schult, and D. Thomson. Dynamic quality-of-service for mobile ad hoc networks. In *MobiHoc '00: Proceedings of the 1st ACM international symposium on Mobile ad hoc networking & computing*, pages 137–138, Piscataway, NJ, USA, 2000. IEEE Press. ISBN 0-7803-6534-8.

[48] Seoung-Bum Lee, Gahng-Seop Ahn, Xiaowei Zhang, and Andrew T. Campbell. INSIGNIA: An IP-based quality of service framework for mobile ad hoc networks. *Journal of Parallel and Distributed Computing*, 60(4):374–406, 2000. URL citeseer.ist.psu.edu/lee00insignia.html.

[49] Prasun Sinha, Raghupathy Sivakumar, and Vaduvur Bharghavan. Cedar: a core-extraction distributed ad hoc routing algorithm. *IEEE INFOCOM*, pages 202–209, 1999.

[50] Shigang Chen and Klara Nahrstedt. A distributed quality-of-service routing in ad-hoc networks. *IEEE Journal on Selected Areas in Communications*, 17(8), August 1999.

[51] Chunhung Richard Lin. On-demand qos routing in multihop mobile networks. In *GLOBECOM '00 IEEE Global Telecommunications Conference*. IEEE, 2000.

[52] Chenxi Zhu and Scott Corson. Qos routing for mobile ad hoc networks. In *Proceedings of the 21st Annual Joint Conference of the IEEE Computer and Communications Societies (INFOCOM)*. IEEE, 2002.

[53] Claude Chaudet and Isabelle Guerin Lassous. Bruit: Bandwidth reservation under interferences influence. In *European Wireless (EW2002)*, 2002.

[54] Isabelle Guerin Lassous Claude Chaudet and Janez Zerovnik. A distributed algorithm for bandwidth allocation in stable ad hoc networks. In *International Conference on Wireless On-Demand Network Systems (WONS)*. IFIP, 2004.

[55] Teresa Tung Rajarshi Gupta, Zhanfeng Jia and Jean Walrand. Interference-aware qos routing (iqrouting) for ad-hoc networks. In *International Conference on Communications (ICC)*. IEEE, 2005.

[56] S.Shah and K. Nahrstedt. Predictive location-based qos routing in mobile ad hoc networks. *IEEE ICC*, April 2002.

[57] Martin Kubisch Emma Carlson and Daniel Hollos. A receiver based protection protocol for wireless multi-hop networks. In *In Proceedings of International Workshop on Performance Evaluation of Wireless Ad Hoc, Sensor, and Ubiquitous Networks (PE-WASUN)*, 2005.

[58] Holger Karl Emma Carlson and Adam Wolisz. Distributed allocation of slots for real-time traffic in a wireless multi-hop network. In *In Proceedings of European Wireless*, 2004.

[59] Sachin Abhyankar Carlos Cordeiro and Dharma Agrawal. Design and implementation of qos-driven dynamic slot assignment and piconet partitioning algorithms over bluetooth wpans. In *INFOCOM'04. Twenty-third AnnualJoint Conference of the IEEE Computer and Communications Societies*. IEEE, 2004.

[60] Shiann-Tsong Sheu and Tzu-Fang Sheu. A bandwidth allocation/sharing/extension protocol for multimedia over ieee 802.11 ad hoc wireless lans. *Selected Areas in Communications*, 19(10), October 2001.

[61] Theodoros Salonidis and Leandros Tassiulas. Distributed dynamic scheduling for end-to-end rate guarantees in wireless ad hoc networks. In *MobiHoc '05: Proceedings of the 6th ACM international symposium on Mobile ad hoc networking and computing*, pages 145–156, New York, NY, USA, 2005. ACM Press. ISBN 1-59593-004-3. doi: http://doi.acm.org/10.1145/1062689.1062709.

[62] Yaling Yang. Contention-aware admission control for ad hoc networks. *Transactions onf Mobile Computing*, 4(4), July 2005.

[63] Yaling Yang and Robin Kravets. Distributed qos guarantees for realtime traffic in ad hoc networks. In *The First IEEE Communications Society Conference on Sensor and Ad Hoc Communications and Networks (SECON)*. IEEE, 2004.

[64] D. C. Cox, R. Murray, and A. Norris. 800 MHz attenuation measured in and around suburban houses. *AT&T Bell Lab. Tech. J.*, 63(6):921–954, 1984.

[65] R. Hekmat and P. Van Mieghem. Connectivity in wireless ad-hoc networks with a log-normal radio model. *Mobile Networks and Applications*, 11:351–360, 2006.

[66] J. Orriss and S. K. Barton. Probability distributions for the number of radio transceivers which can communicate with one another. *IEEE Trans. Commun.*, 51(4):676–681, 2003.

[67] D. Miorandi and E. Altman. Coverage and connectivity of ad hoc networks in presence of channel randomness. In *Proc. IEEE 24th Annual Joint Conference of the IEEE Computer and Communications Societies (INFOCOM)*, pages 491–502, March 2005.

[68] M. D. Penrose. On the spread-out limit for bond and continuum percolation. *Ann. Appl. Probab.*, 3(1):253–276, 1993.

[69] L. Booth, J. Bruck, M. Cook, and M. Franceschetti. Ad hoc wireless networks with noisy links. In *Proc. IEEE International Symposium on Information Theory (ISIT)*, page 386, Yokohama, Japan, June 2003.

[70] R. Meester and R. Roy. *Continuum Percolation*. Cambridge University Press, 1996.

[71] Torsten Muetze, Patrick Stuedi, Fabian Kuhn, and Gustavo Alonso. Understanding radio irregularity in wireless networks, 2008. IEEE Secon.

[72] The VINT Project. The NS network simulator, 2002. URL `Available at http://www-mash.CS.Berkeley.EDU/ns`.

[73] Sven Krumke and Madhav Marathe. Models and approximation algorithms for channel assignment in radio networks. *Wirel. Netw.*, 7(6):575–584, 2001. ISSN 1022-0038. doi: http://dx.doi.org/10.1023/A:1012311216333.

[74] T.H. Cormen, C.E. Leiserson, and R.L. Rivest. *Introduction to Algorithms*. MIT Press, 1990.

[75] Himabindu Puch, Saumitra Das, and Charlu Hu. Ekta: An efficient dht substrate for distributed applications in mobile ad hoc networks, 2004. Proceedings of the 6th IEEE Workshop on Mobile Computing Systems and Applications (WMCSA 2004).

[76] Filipe Araujo, Luis Rodrigues, Joerg Kaiser, and Changling Liu. Chr: a distributed hash table for wireless ad hoc networks, 2005. Proceedings of the 25th IEEE International Workshop on Distributed Computing Systems.

[77] Thomas Zahn and Jochen Schiller. Madpastry: A dht substrate for practicably sized manets, 2005. Proceedings of the 5th Workshop on Applications and Services in Wireless Networks (ASWN 2005).

[78] Curt Cramer and Thomas Fuhrmann. Proximity neighbor selection for a dht in wireless multi-hop networks. In *Peer-to-Peer Computing*, pages 3–10, 2005.

[79] Raphaël Kummer, Peter Kropf, and Pascal Felber. Distributed lookup in structured peer-to-peer ad-hoc networks. In *OTM Conferences (2)*, pages 1541–1554, 2006.

[80] Specification of the Bluetooth System, Core Package version 1.2, November 2003. https://www.bluetooth.org/spec/.

[81] Wireless Medium Access Control (MAC) and Physical Layer (PHY) Specifications for Low-Rate Wireless Personal Area Networks (LR-WPANs), October 2003. http://standards.ieee.org/getieee802/download/802.15.4-2003.pdf.

[82] Wireless LAN Medium Access Control (MAC) and Physical Layer (PHY) Specifications, June 2003. http://standards.ieee.org/getieee802/download/802.11-1999.pdf.

[83] R. Stewart, Q. Xie, K. Morneault, C. Sharp, H. Schwarzbauer, T. Taylor, I. Rytina, M. Kalla, L. Zhang, and V. Paxson. Stream Control Transmission Protocol. RFC 2960 (Proposed Standard), October 2000. URL http://www.ietf.org/rfc/rfc2960.txt. Updated by RFC 3309.

[84] Ad-hoc On-demand Distance Vector Routing – Uppsala Implementation, 2007. http://core.it.uu.se/core/index.php/AODV-UU.

[85] M. J. Chang, M. J. Lee, and S. J. Koh. Address management for mobile sctp handover. Internet Draft, October 2004. URL http://www.ietf.org/internet-drafts/draft-chang-mobile-sctp-address-mgt-01.

[86] Andreas Frei and Gustavo Alonso. A dynamic lightweight architecture. *Proceedings of the 3rd International Conference on Pervasive Computing and Communications (PerCom)*, 2005.

[87] JXTA v2.0 Protocols Specification, October 2003. http://spec.jxta.org/nonav/-v1.0/docbook/JXTAProtocols.html.

[88] Daniel Brookshier, Darren Govoni, and Navaneeth Krishnan. *JXTA: Java P2P Programming*. SAMS, March 2002.

[89] Media Access Control (MAC) Bridges, June 2004. http://standards.ieee.org/getieee802/802.1.html.

[90] Linux Ethernet Bridge, 2008. http://bridge.sourceforge.net.

[91] Bluetooth Network Encapsulation Protocol, June 2001. http://grouper.ieee.org/groups/802/15/Bluetooth/BNEP.pdf.

[92] Bluez, Official Linux Bluetooth protocol stack, 2006. http://www.bluez.org.

[93] C. Perkins, E. Belding-Royer, and S. Das. Ad hoc On-Demand Distance Vector (AODV) Routing. RFC 3561 (Experimental), July 2003. URL http://www.ietf.org/rfc/rfc3561.txt.

[94] T. Clausen, P. Jacquet (editors), C. Adjih, A. Laouiti, P. Minet, P. Muhlethaler, A. Qayyum, and L.Viennot. Optimized link state routing protocol (olsr). RFC 3626, October 2003. URL http://ietf.org/rfc/rfc3626.txt. Network Working Group.

[95] OLSRD - An ad hoc wireless mesh routing daemon, 2007. http://www.olsr.org/.

[96] C. Perkins and E. Royer. Ad hoc on-demand distance vector routing. *Proceedings of the 2nd IEEE Workshop on Mobile Computing Systems and Applications*, pages 90–100, Feb 1999.

[97] The Iperf Project, 2003. http://dast.nlanr.net/Projects/Iperf.

[98] Chandrakanth Chereddi, Pradeep Kyasanur, and Nitin H. Vaidya. Design and implementation of a multi-channel multi-interface network. In *REALMAN '06: Proceedings of the 2nd international workshop on Multi-hop ad hoc networks: from theory to reality*, pages 23–30, New York, NY, USA, 2006. ACM. ISBN 1-59593-360-3. doi: http://doi.acm.org/10.1145/1132983.1132988.

[99] Chandrakanth Chereddi, Pradeep Kyasanur, and Nitin H. Vaidya. Net-x: a multichannel multi-interface wireless mesh implementation. *SIGMOBILE Mob. Comput. Commun. Rev.*, 11(3):84–95, 2007. ISSN 1559-1662. doi: http://doi.acm.org/10.1145/1317425.1317435.

[100] R. Chandra and P. Bahl. Multinet: connecting to multiple ieee 802.11 networks using a single wireless card. *INFOCOM 2004. Twenty-third AnnualJoint Conference of the IEEE Computer and Communications Societies*, 2:882–893 vol.2, March 2004. ISSN 0743-166X.

[101] Virtual Wifi. http://research.microsoft.com/netres/projects/virtualwifi.

[102] N. Boulicault, G. Chelius, and E. Fleury. Ana4: a 2.5 framework for deploying real multi-hop ad hoc and mesh networks. *Ad Hoc and Sensor Wireless Networks: an International Journal (AHSWN)*, to be published, 2005.

[103] V. Kawadia, Yongguang Zhang, and B. Gupta. System services for implementing ad-hoc routing protocols. *Parallel Processing Workshops, 2002. Proceedings. International Conference on*, pages 135–142, 2002. ISSN 1530-2016. doi: 10.1109/ICPPW.2002.1039723.

[104] Richard Draves, Jitendra Padhye, and Brian Zill. Routing in multi-radio, multi-hop wireless mesh networks. In *MobiCom '04: Proceedings of the 10th annual international conference on Mobile computing and networking*, pages 114–128. ACM Press, 2004. ISBN 1581138687. doi: http://dx.doi.org/10.1145/1023720.1023732. URL http://dx.doi.org/10.1145/1023720.1023732.

[105] Richard Draves, Jitendra Padhye, and Brian Zill. Comparison of routing metrics for static multi-hop wireless networks. In *SIGCOMM '04: Proceedings of the 2004 conference on Applications, technologies, architectures, and protocols for computer communications*, volume 34, pages 133–144, New York, NY, USA, October 2004. ACM Press. ISBN 1581138628. doi: http://dx.doi.org/10.1145/1015467.1015483. URL http://dx.doi.org/10.1145/1015467.1015483.

[106] Mobile Unlimited, 2006. http://www.swisscom.ch/solutions/mobile-unlimited.

[107] Mo Li, Kumbesan Sandrasegaran, and Tracy Tung. A multi-interface proposal for ieee 802.21 media independent handover. In *ICMB '07: Proceedings of the International Conference on the Management of Mobile Business*, page 7, Washington, DC, USA, 2007. IEEE Computer Society. ISBN 0-7695-2803-1. doi: http://dx.doi.org/10.1109/ICMB.2007.2.

[108] Ashutosh Dutta, Subir Das, David Famolari, Yoshihiro Ohba, Kenichi Taniuchi, Victor Fajardo, Rafa Marin Lopez, Toshikazu Kodama, and Henning Schulzrinne. Seamless proactive handover across heterogeneous access networks. *Wirel. Pers. Commun.*, 43(3):837–855, 2007. ISSN 0929-6212. doi: http://dx.doi.org/10.1007/s11277-007-9266-3.

[109] Li Ma, Fei Yu, V.C.M. Leung, and T. Randhawa. A new method to support umts/wlan vertical handover using sctp. *Wireless Communications, IEEE*, 11 (4):44–51, Aug. 2004. ISSN 1536-1284. doi: 10.1109/MWC.2004.1325890.

[110] U. Javaid, D.-E. Meddour, T. Rasheed, and T. Ahmed. Personal network routing protocol (pnrp) for personal ubiquitous environments. *Communications, 2007. ICC '07. IEEE International Conference on*, pages 3100–3107, June 2007. doi: 10.1109/ICC.2007.515.

[111] K. Sethom, M. Sabeur, B. Jouaber, H. Afifi, and D. Zeghlache. Distributed virtual network interfaces to support intra-pan and pan-to-infrastructure connectivity. *Global Telecommunications Conference, 2005. GLOBECOM '05. IEEE*, 6:5 pp.–, Nov.-2 Dec. 2005. doi: 10.1109/GLOCOM.2005.1578434.

[112] Seth Gilbert and Nancy Lynch. Brewers conjecture and the feasibility of consistent, available, partition-tolerant web services. *Sigact News*, 33:51–59, 2002.

[113] The Netfilter/Iptables Project, June 2001. http://www.netfilter.org.

[114] T. Clausen and P. Jacquet. Optimized Link State Routing Protocol (OLSR). RFC 3626 (Experimental), October 2003. URL http://www.ietf.org/rfc/rfc3626.txt.

[115] Francisco J. Ros and Pedro M. Ruiz. Implementing a new manet unicast routing protocol in ns2. Technical report, Universidad de Murcia, 2004. URL http://sourceforge.net/projects/um-olsr.

[116] Christoph Lindemann and Oliver P. Waldhorst. Exploiting epidemic data dissemination for consistent lookup operations in mobile applications. *SIGMOBILE Mob. Comput. Commun. Rev.*, 8(3):44–56, 2004. ISSN 1559-1662. doi: http://doi.acm.org/10.1145/1031483.1031490.

[117] Christoph Lindemann and Oliver P. Waldhorst. Modeling epidemic information dissemination on mobile devices with finite buffers. *SIGMETRICS Perform. Eval. Rev.*, 33(1):121–132, 2005. ISSN 0163-5999. doi: http://doi.acm.org/10.1145/1071690.1064227.

[118] Curt Cramer and Thomas Fuhrmann. Proximity neighbor selection for a dht in wireless multi-hop networks. In *P2P '05: Proceedings of the Fifth IEEE*

International Conference on Peer-to-Peer Computing, pages 3–10, Washington, DC, USA, 2005. IEEE Computer Society. ISBN 0-7695-2376-5. doi: http://dx.doi.org/10.1109/P2P.2005.28.

[119] Ion Stoica, Robert Morris, David Liben-Nowell, David Karger, M. Frans Kaashoek, Frank Dabek, and Hari Balakrishnan. Chord: A scalable peer-to-peer lookup service for internet applications. *IEEE Transactions on Networking*, 11, February 2003.

[120] Y. Charlie Hu, Saumitra M. Das, and Himabindu Pucha. Exploiting the synergy between peer-to-peer and mobile ad hoc networks. In *HOTOS'03: Proceedings of the 9th conference on Hot Topics in Operating Systems*, pages 7–7, Berkeley, CA, USA, 2003. USENIX Association.

[121] Antony I. T. Rowstron and Peter Druschel. Pastry: Scalable, decentralized object location, and routing for large-scale peer-to-peer systems. In *Middleware '01: Proceedings of the IFIP/ACM International Conference on Distributed Systems Platforms Heidelberg*, pages 329–350, London, UK, 2001. Springer-Verlag. ISBN 3-540-42800-3.

[122] David B Johnson and David A Maltz. Dynamic source routing in ad hoc wireless networks. In Imielinski and Korth, editors, *Mobile Computing*, volume 353. Kluwer Academic Publishers, 1996. URL citeseer.ist.psu.edu/johnson96dynamic.html.

[123] Matthew Caesar, Miguel Castro, Edmund B. Nightingale, Greg O'Shea, and Antony Rowstron. Virtual ring routing: network routing inspired by dhts. *SIGCOMM Comput. Commun. Rev.*, 36(4):351–362, 2006. ISSN 0146-4833. doi: http://doi.acm.org/10.1145/1151659.1159954.

[124] Francoise Sailhan and Valerie Issarny. Scalable service discovery for manet. In *Proceedings of the 3rd IEEE international Conference on Pervasive Computing and Communications (PerCom'05)*. IEEE, 2005.

[125] P.E. Engelstad and Y. Zheng. Evaluation of service discovery architectures for mobile ad hoc network. In *WONS '05: Proceedings of the Second Annual Conference on Wireless On-demand Network Systems and Services*. IEEE, 2005.

[126] J. Eriksson, M. Faloutsos, and S. Krishnamurthy. Scalable ad hoc routing: the case for dynamic addressing. *INFOCOM 2004. Twenty-third AnnualJoint Conference of the IEEE Computer and Communications Societies*, 2:1108–1119 vol.2, March 2004. ISSN 0743-166X.

[127] Ananth Rao, Sylvia Ratnasamy, Christos Papadimitriou, Scott Shenker, and Ion Stoica. Geographic routing without location information. In *MobiCom '03: Proceedings of the 9th annual international conference on Mobile computing and networking*, pages 96–108, New York, NY, USA, 2003. ACM. ISBN 1-58113-753-2. doi: http://doi.acm.org/10.1145/938985.938996.

[128] Aline C. Viana, Marcelo D. de Amorim, Serge Fdida, Yannis Viniotis, and José F. de Rezende. Easily-managed and topology-independent location service for self-organizing networks. In *MobiHoc '05: Proceedings of the 6th*

ACM international symposium on Mobile ad hoc networking and computing, pages 193–204, New York, NY, USA, 2005. ACM. ISBN 1-59593-004-3. doi: http://doi.acm.org/10.1145/1062689.1062714.

[129] Jinyang Li, John Jannotti, Douglas S. J. De Couto, David R. Karger, and Robert Morris. A scalable location service for geographic ad hoc routing. In *MobiCom '00: Proceedings of the 6th annual international conference on Mobile computing and networking*, pages 120–130, New York, NY, USA, 2000. ACM. ISBN 1-58113-197-6. doi: http://doi.acm.org/10.1145/345910.345931.

[130] Yuan Xue, Baochun Li, and Klara Nahrstedt. A scalable location management scheme in mobile ad-hoc networks. In *LCN '01: Proceedings of the 26th Annual IEEE Conference on Local Computer Networks*, page 102, Washington, DC, USA, 2001. IEEE Computer Society.

[131] Yinzhe Yu, Guor-Huar Lu, and Zhi-Li Zhang. Enhancing location service scalability with high-grade. *Mobile Ad-hoc and Sensor Systems, 2004 IEEE International Conference on*, pages 164–173, Oct. 2004.

[132] Ittai Abraham. Lls : a locality aware location service. In *In Proceedings of the DIALM-POMC Joint Workshop on Foundations of Mobile Computing (DIALM-POMC 2004*, 2004.

[133] E. Guttman, C. Perkins, J. Veizades, and M. Day. Service Location Protocol, Version 2 . RFC 2608 (Proposed Standard), June 1999. URL http://www.ietf.org/rfc/rfc2608.txt. Updated by RFC 3224.

[134] Jini Specification, 1999. http://java.sun.com/jini.

[135] Universal Description Discovery and Integration Platform, 2000. http://www.uddi.org.

[136] Salutation Consortium: Salutation Architecture Specification, 1999, 2006. http://www.salutation.org/specordr.htm.

[137] Microsoft Corporation, Universal Plug and Play, 1999. http://www.upnp.org/resources.

[138] Specification of the Bluetooth System, 1999. http://www.bluetooth.com.

[139] U.C. Kozat and L. Tassiulas. Network layer support for service discovery in mobile ad hoc networks. *INFOCOM 2003. Twenty-Second Annual Joint Conference of the IEEE Computer and Communications Societies. IEEE*, 3:1965–1975 vol.3, March-3 April 2003. ISSN 0743-166X.

[140] Liang Cheng. Service advertisement and discovery in mobile ad hoc networks. In *In Proc. of CSCW 2002*, 2002.

[141] Sung Ju Lee, William Su, and Mario Gerla. On-demand multicast routing protocol in multihop wireless mobile networks. *Mob. Netw. Appl.*, 7(6):441–453, 2002. ISSN 1383-469X. doi: http://dx.doi.org/10.1023/A:1020756600187.

[142] Jose Luis Jodra, Maribel Vara, Jose Ma Cabero, and Josu Bagazgoitia. Service discovery mechanism over olsr for mobile ad-hoc networks. In *AINA '06: Proceedings of the 20th International Conference on Advanced Information Networking and Applications - Volume 2 (AINA '06)*, pages 534–542, Washington, DC, USA, 2006. IEEE Computer Society. ISBN 0-7695-2466-4-02. doi: http://dx.doi.org/10.1109/AINA.2006.305.

[143] Li Li and Louise Lamont. A lightweight service discovery mechanism for mobile ad hoc pervasive environment using cross-layer design. In *PERCOMW '05: Proceedings of the Third IEEE International Conference on Pervasive Computing and Communications Workshops*, pages 55–59, Washington, DC, USA, 2005. IEEE Computer Society. ISBN 0-7695-2300-5. doi: http://dx.doi.org/10.1109/PERCOMW.2005.8.

[144] R. Koodli and C.E Perkins. Service Discovery in On-Demand Ad Hoc Networks . Internet Draft, October 2002. URL draft-koodli-manet-servicediscovery-00.txt. Work in progress.

[145] Anders Nilsson, Charles E. Perkins, Antti J. Tuominen, Ryuji Wakikawa, and Jari T. Malinen. Aodv and ipv6 internet access for ad hoc networks. *SIGMOBILE Mob. Comput. Commun. Rev.*, 6(3):102–103, 2002. ISSN 1559-1662. doi: http://doi.acm.org/10.1145/581291.581310.

[146] Habib Ammari and Hesham El-Rewini. Integration of mobile ad hoc networks and the internet using mobile gateways. In *Proceedings of 18th International Parallel and Distributed Processing Symposium*, volume 13, page 218b, Los Alamitos, CA, USA, 2004. IEEE Computer Society. ISBN 0-7695-2132-0. doi: http://doi.ieeecomputersociety.org/10.1109/IPDPS.2004.1303253.

[147] Yuan Sun, Elizabeth M. Belding-Royer, and Charles E. Perkins. Internet connectivity for ad hoc mobile networks. *International Journal of Wireless Information Networks special issue on "Mobile Ad Hoc Networks (MANETs): Standards, Research, Applications"*, 9(2), April 2002.

[148] Prashant Ratanchandani and Robin Kravets. A hybrid approach to internet connectivity for mobile ad hoc networks. In *Proceedings of IEEE WCNC*, 2003.

[149] Marcin Michalak and Torsten Braun. Common gateway architecture for mobile ad-hoc networks. In *WONS'05: Second Annual Conference on Wireless On demand Network Systems and Services*, 2005.

[150] Ulf Jönsson, Fredrik Alriksson, Tony Larsson, Per Johansson, and Jr. Gerald Q. Maguire. Mipmanet: mobile ip for mobile ad hoc networks. In *MobiHoc '00: Proceedings of the 1st ACM international symposium on Mobile ad hoc networking & computing*, pages 75–85, Piscataway, NJ, USA, 2000. IEEE Press. ISBN 0-7803-6534-8.

[151] Aaron Striegel, Ranga S. Ramanujan, and Jordan Bonney. A protocol independent internet gateway for ad hoc wireless networks. In *Proceedings of 26th annual conference on Local Computer Networks (LCN)*, volume 00, page 92,

Los Alamitos, CA, USA, 2001. IEEE Computer Society. ISBN 0-7695-1321-2. doi: http://doi.ieeecomputersociety.org/10.1109/LCN.2001.990774.

[152] Ryuji Wakikawa, Jari T. Malinen, Charles E. Perkins, Anders Nilsson, and Antti J. Tuominen. Global connectivity for ipv6 mobile ad hoc networks. Internet Draft, March 2006. URL http://www.ietf.org/internet-drafts/draft-wakikawa-manet-globalv6-05.txt.

[153] J. Rosenberg, J. Weinberger, C. Huitema, and R. Mahy. STUN - Simple Traversal of User Datagram Protocol (UDP) Through Network Address Translators (NATs). RFC 3489 (Proposed Standard), March 2003. URL http://www.ietf.org/rfc/rfc3489.txt.

[154] P.V. Mockapetris. Domain names - implementation and specification. RFC 1035 (Standard), November 1987. URL http://www.ietf.org/rfc/rfc1035.txt. Updated by RFCs 1101, 1183, 1348, 1876, 1982, 1995, 1996, 2065, 2136, 2181, 2137, 2308, 2535, 2845, 3425, 3658, 4033, 4034, 4035, 4343.

[155] Paal Engelstad, Do Van Thanh, and Tore Jonvik. Name resolution in mobile ad-hoc networks. In *ICT'03: International Conference on Telecommunications*, 2003.

[156] Paal Engelstad, Do Van Thanh, and Geir Egeland. Name resolution in mobile ad-hoc networks and over external ip networks. In *ICC'03: International Conference on Communications*, 2003.

[157] Xiaoyan Hong, Jun Liu, Randy Smith, and Yeng-Zhong Lee. Distributed naming system for mobile ad hoc network. In *ICWN*, pages 509–515, 2005.

[158] Christophe Jelger and Christian Tschudin. Model based protocol fusion for manet-internet integration. In *WONS'06: Proceeding of the 3rd Annual Conference on Wireless On demand Network Systems and Services*, 2006.

[159] Christophe Jelger and Christian Tschudin. Underlay fusion of dns, arp/nd, and path resolution in manets. In *ADHOC'05: Proceedings of the 5th Scandinavian Workshop on Wireless Ad-hoc Networks*, 2005.

[160] Jaehoon Jeong, Jungsoo Park, and Hyoungjun Kim. Name directory service based on maodv and multicast dns for ipv6 manet. *Vehicular Technology Conference, 2004. VTC2004-Fall. 2004 IEEE 60th*, 7:4750–4753 Vol. 7, 26-29 Sept. 2004. ISSN 1090-3038. doi: 10.1109/VETECF.2004.1404994.

[161] Jaehoon Jeong, Jungsoo Park, and Hyoungjun Kim. Name service in ipv6 mobile ad-hoc network connected to the internet. *Personal, Indoor and Mobile Radio Communications, 2003. PIMRC 2003. 14th IEEE Proceedings on*, 2: 1351–1355 vol.2, 7-10 Sept. 2003. doi: 10.1109/PIMRC.2003.1260333.

[162] Christophe Jelger and Christian Tschudin. Model based protocol fusion for manet-internet integration. In *WONS'06: Proceedings of the 3rd Annual Conference on Wireless On demand Network Systems and Services*, 2006.

[163] Mick O'Doherty. Pico sip. Internet Draft, February 2001.

[164] Simone Leggio, Jukka Manner, Antti Hulkkonen, and Kimmo Raatikainen. Session initiation protocol deployment in ad-hoc networks: a decentralized approach. In *2nd International Workshop on Wireless Ad-hoc Networks (IWWAN)*, 2005.

[165] Hyun-Gon Seo amd Ki-Hyung Kim, Won-Do Jung, and Jung-Sung Park. Performance of service location protocols in manet based on reactive routing protocols. In *Proceedings of 4th International Conference on Networking (ICN'05)*, 2005.

[166] Li Li and Louise Lamont. A lightweight service discovery mechanism for mobile ad hoc pervasive environment using cross-layer design. In *Percom'05: Proceeding of the 3rd International Conference on Pervasive Computing and Communications Workshops*, 2005.

[167] A. Dutta, R. Jain, K. Wong, J. Burns, K. Young, and H. Schulzrinne. Multilayered mobility management for survivable network. In *Proceedings of MILCOM*, 2001.

[168] K. Egevang and P. Francis. The IP Network Address Translator (NAT). RFC 1631 (Informational), May 1994. URL http://www.ietf.org/rfc/rfc1631.txt. Obsoleted by RFC 3022.

[169] Jen-Jee Chen, Yu-Li Cheng, Yu-Chee Tseng, and Quincy Wu. A push-based voip service for an internet-enabled mobile ad hoc network. In *Proceedings of 3rd IEEE VTS Asia Pacific Wireless Comm. Symposium (APWCS'06)*, 2006.

[170] Matthias Grossglauser and David N. C. Tse. Mobility increases the capacity of ad hoc wireless networks. *IEEE/ACM Trans. Netw.*, 10(4):477–486, 2002. ISSN 1063-6692. doi: http://dx.doi.org/10.1109/TNET.2002.801403.

[171] Christian Plattner. *Ganymed: A Platform for Database Replication*. PhD thesis, Swiss Federal Institute of Technology Zurich, 2006.

[172] Open extension to CGI, 2008. http://www.fastcgi.com.

[173] The Nokia Internet Tables, 2007. http://www.nokia.com/n800.

[174] Maemo Linux, 2008. http://www.maemo.org.

[175] OpenSocial: Common application programming interfaces for web-based social network applications, . http://code.google.com/apis/opensocial.

[176] OpenID: An open and decentralized identity system, . http://openid.net.

[177] Elizabeth M. Daly and Mads Haahr. Social network analysis for routing in disconnected delay-tolerant manets. In *MobiHoc '07: Proceedings of the 8th ACM international symposium on Mobile ad hoc networking and computing*, pages 32–40, New York, NY, USA, 2007. ACM. ISBN 978-1-59593-684-4. doi: http://doi.acm.org/10.1145/1288107.1288113.

[178] Klaus Herrmann. Modeling the sociological aspects of mobility in ad hoc networks. In *MSWIM '03: Proceedings of the 6th ACM international workshop on Modeling analysis and simulation of wireless and mobile systems*, pages 128–129, New York, NY, USA, 2003. ACM. ISBN 1-58113-766-4. doi: http://doi.acm.org/10.1145/940991.941014.

[179] Desmond J. Higham. Unravelling small world networks. *J. Comput. Appl. Math.*, 158(1):61–74, 2003. ISSN 0377-0427. doi: http://dx.doi.org/10.1016/S0377-0427(03)00471-0.

[180] Stephen Wicker Shankar Sastry Sameer Pai, Tanya Roosta. Using social network theory towards development of wireless ad hoc network trust. In *Proceedings of the IEEE 21st International Conference on Advanced Information Networking and Applications*, June 2007. URL http://www.truststc.org/pubs/202.html.

[181] Mobikade: mobile social networking service. http://www2.mkade.com.

[182] Dodgeball: Mobile Social Software. http://www.dodgeball.com.

Die VDM Verlagsservicegesellschaft sucht für wissenschaftliche Verlage abgeschlossene und herausragende

Dissertationen, Habilitationen, Diplomarbeiten, Master Theses, Magisterarbeiten usw.

für die kostenlose Publikation als Fachbuch.

Sie verfügen über eine Arbeit, die hohen inhaltlichen und formalen Ansprüchen genügt, und haben Interesse an einer honorarvergüteten Publikation?

Dann senden Sie bitte erste Informationen über sich und Ihre Arbeit per Email an *info@vdm-vsg.de*.

Sie erhalten kurzfristig unser Feedback!

VDM Verlagsservicegesellschaft mbH
Dudweiler Landstr. 99　　　　　　Telefon +49 681 3720 174
D - 66123 Saarbrücken　　　　　　Fax   +49 681 3720 1749
www.vdm-vsg.de

Die VDM Verlagsservicegesellschaft mbH vertritt

Printed by Books on Demand GmbH, Norderstedt / Germany